P9-DVL-524

Direct Broadcast Satellite Communications

ISBN 0-201-69582-0

The Addison-Wesley Wireless Communications Series
Andrew J. Viterbi, Consulting Editor

Wireless Personal Communications Systems
David J. Goodman

Direct Broadcast Satellite Communications: An MPEG Enabled Service
Donald C. Mead

Mobile IP: Design Principles and Practices
Charles E. Perkins

CDMA: Principles of Spread Spectrum Communication
Andrew J. Viterbi

Wireless Multimedia Communications: Networking Video, Voice, and Data
Ellen Kayata Wesel

Direct Broadcast Satellite Communications

An MPEG Enabled Service

Donald C. Mead

Addison-Wesley Wireless Communications Series
A Prentice Hall Title
Upper Saddle River, NJ 07458
http://www.phptr.com

Many of the designations used by manufacturers and sellers to distinguish their products are claimed as trademarks. Where those designations appear in this book and Prentice Hall was aware of a trademark claim, the designations have been printed in initial caps or all caps.

The author and publisher have taken care in the preparation of this book, but make no expressed or implied warranty of any kind and assume no responsibility for errors or omissions. No liability is assumed for incidental or consequential damages in connection with or arising out of the use of the information or programs contained herein.

The publisher offers discounts on this book when ordered in quantity for special sales. For more information, please contact: Corporate Sales Department, Phone: 800-382-3419, Fax: 201-236-7141 E-mail: corpsales@prenhall.com, or write: Prentice Hall PTR, Corporate Sales Department, One Lake Street, Upper Saddle River, NJ 07458

Library of Congress Cataloging-in-Publication Data

Mead, Donald C.
Direct broadcast satellite communications : an MPEG enabled service / Donald C. Mead.
p. c.m. — (The Addison-Wesley wireless communications series)
Includes bibliographical references and index.
ISBN 0-201-69582-0
1. Direct broadcast satellite television. 2. MPEG (Video coding standard) I. Title. II. Series.
TK6677.M43 1999
621.388′53--dc21 98-56157
CIP

Copyright © 2000 by Prentice Hall PTR
Prentice -Hall, Inc.
Upper Saddle River, NJ 07458

All rights reserved. No part of this publication may be reproduced, stored in a retrieval system, or transmitted, in any form, or by any means, electronic, mechanical, photocopying, recording, or otherwise, without the prior consent of the publisher.
Printed in the United States of America.

ISBN 0-201-69582-0

10 9 8 7 6 5 4 3 2 1

Prentice-Hall International (UK) Limited, London
Prentice-Hall of Australia Pty. Limited, Sydney
Prentice-Hall Canada Inc., Toronto
Prentice-Hall Hispanoamericana, S.A., Mexico
Prentice-Hall of India Private Limited, New Delhi
Prentice-Hall of Japan, Inc., Tokyo
Prentice-Hall (Singapore) Pte. Ltd., Singapore
Editora Prentice-Hall do Brasil, Ltda., Rio de Janeiro

CONTENTS

FOREWORD

The direct broadcast satellites that currently transmit hundreds of video channels into very small dishes are direct descendants of the pioneering SYNCOM satellites developed more than 35 years ago. Although severely limited in power, antenna gain, and bandwidth, the first of these to operate in a geostationary orbit, SYNCOM 3, was nonetheless able to relay a single black-and-white television signal across the Pacific Ocean, transmitting the Tokyo Olympics to the United States in 1964. The extraordinary receiving sensitivity that accomplished this tour de force was achieved by using a very large earth terminal equipped with a liquid-helium-cooled maser, a combination not exactly suited for use in the average home.

During the succeeding years, increases in the transmitter power and antenna gain of the satellites has resulted in more than a millionfold increase in effective radiated power. Improvements in solid-state technology have resulted in uncooled, low-cost ground station receivers nearly as sensitive as the maser. This combination allows the use of the small dishes for television reception, but the limited bandwidth allocated for direct broadcast service limits the number of uncompressed analog television signals to a number too small to create a profitable commercial service.

The final technological advance needed to make a business out of direct broadcast satellites was the development of digital compression techniques and standards. Digital compression multiplies the number of television signals that can be provided by a substantial factor while improving the picture quality. The very low-cost decompressor used in the home equipment required the use of state-of-the-art digital design and manufacturing techniques.

The DBS satellites represent a significant milestone in the development of communications. Don Mead was deeply involved in the creation of geostationary communication satellites, having been responsible for the digital circuits used in SYNCOM's telemetry and command systems. When Hughes started what became DIRECTV in 1990, Don became our compression guru, creating a laboratory in which the candidate systems could be compared, and serving on the international committee that selected the detailed international standard.

I am sure you will find this book stimulating and instructive because it covers in one place all of the disciplines involved in DBS communications.

Harold A. Rosen

PREFACE

This book, intended for electronics and communications engineers, describes how all of the individual developments of today's Direct Broadcast Satellites (DBS) came together to provide an overall communication system capable of delivering more than 200 audio/video services.

The state of the art in communications technology is changing so rapidly that it is difficult for anyone associated with electronic communications to remain current. The developments in compression, in particular, are proceeding at a pace that exceeds even the staggering rate of Moore's law, which predicts the increasing capabilities of semiconductors that underlie almost all current technologies.

This book starts with a specific communication system, DBS services, and then shows how the MPEG 1 and MPEG 2 standards were used to implement this system. Thus, the book provides the reader with not just an MPEG or communications satellite discussion, but a complete discussion of how the MPEG standards are used to implement a modern satellite broadcast system.

Organization of the Book

The book is divided into the following five parts:

Part One provides an overview of DBS. This includes Chapter 1, History of Communication Satellites; Chapter 2, Regulatory Framework, including international and Federal Communications Commission regulations; and Chapter 3, An Overview of the DBS System.

Part Two describes the key subsystems and design decisions for DBS. These include Chapter 4, Key Elements of the Radio Frequency Subsystem; Chapter 5, Forward Error Correction; and Chapter 6, Conditional Access.

Part Three describes the key elements of the MPEG international standards as they apply to DBS. It includes Chapter 7, MPEG 2 Systems; Chapter 8, MPEG 2 Video Compression; and Chapter 9, MPEG 1 Audio Compression.

Part Four describes the ground subsystems that connect the customer to the satellite: Chapter 10, DBS Uplink Facilities; and Chapter 11, Integrated Receiver Decoder.

Part Five explores some future digital satellite services and technologies. These include Chapter 12, Spaceway and the Global Broadcast Service; and Chapter 13, Intelligent Compression: MPEG 4.

Using This Book

This book is intended for a diverse group of readers, ranging from those who want to obtain a general overview of Direct Broadcast Satellites to those who want to delve deeply into one or all of the technical facets of DBS systems. To accommodate this diversity, sections within the book are annotated by a marginal icon system:

- No icon means the material is suitable for all readers.

- The rectangular satellite icon means the section contains some technically difficult material.

- A circular world icon means the section contains serious technical material and probably should only be read by those desiring to gain in-depth knowledge of the subject.

Certain reference materials, which make the book more self-contained for communications engineers, are included in the appendices. Technical decisions made by DIRECTV™ and the international Digital Video Broadcast standard are used as case studies throughout the book.

Acknowledgments

No book of this nature could possibly be written without the support of a very large number of people. First, all of DBS would have never been possible without Dr. Harold Rosen, my mentor and friend of many years. Dr. George Hrycenko of Hughes Electronics and Suzanne Holdings of the FCC both helped enormously with reference materials for Chapter 2.

Another major contributor to DBS is MPEG, so I want to acknowledge the efforts of all of my MPEG colleagues and, in particular, those of the "indispensable man"—the convenor of MPEG, Dr. Leonardo Chiariglione.

Nancy Nilson assisted in the typing of the manuscript. David Dunnmeyer was our faithful proofreader, and my wife, Barbara, took care of many of the administrative efforts. Irv Rabowsky, my long-time colleague, was a constant source of encouragement and help with a number of aspects of the book, and he personally reviewed Chapter 9.

Simon Yates of Thomson Publishing originated the concept of the book, and Karen Gettman, Mary Hart, and Tara Herries of Addison Wesley Longman gave constant encouragement and support during the writing. I also wish to thank the other AWL employees for their support throughout.

Finally, I dedicate the book to the memory of my mother, Pearl Marie Mead, who was my biggest fan. She looked forward to the publication of this book, but her untimely death came before she could enjoy its completion.

PART ONE

Overview

CHAPTER ONE

History of Communication Satellites

In the United States today, powerful direct broadcast satellites (DBS) transmit more than 200 audio/video services directly into homes through small, 18-inch-diameter antennas. In the next few years, the number of services surely will exceed 500 and coverage will be extended to many more countries.

The powerful satellites that provide these services are the result of a natural evolution that includes:

- Visionary concepts
- Engineering genius
- Technology continuing to build on itself

This chapter describes the history of this evolution.

1.1 Background

From the dawn of electronic communication in the early twentieth century until the advent of the communication satellite, long-distance communication was limited at best. The electromagnetic waves that implement electronic communication travel in a straight line; the Earth is curved and cannot be penetrated by these waves. Thus, relays in one form or another are required if points wishing to communicate are more than a few hundred miles apart.

One such relay technique is to bounce the signal off the Earth's *ionosphere*—a shell of charged particles that surrounds the Earth and reflects the longer wavelength radio waves back to Earth. Since the frequency of an electromagnetic signal and its wavelength are inversely related, lower-frequency signals are reflected. Even long wavelength communications are erratic and subject to outages because of atmospheric conditions.

Perhaps nothing emphasizes the lack of reliability of previous relay techniques, and the problems such a lack of reliability causes, than the story of what happened on the morning of December 7, 1941. U.S. codebreakers in Washington, DC, had broken the Japanese codes and were sure that an attack on the U.S. Naval base in Honolulu, Hawaii, was imminent. They tried to warn the armed forces there by using radio circuits from San Francisco to Hawaii; however, they were down because of atmospheric conditions. As we are aware from history, the message sent to attempt to warn U.S. forces using commercial Western Union services arrived too late.

The technique of reflecting electromagnetic waves off of the ionosphere does not work at all for the shorter wavelengths that pass through the ionosphere and, consequently, are not reflected back to Earth. Since the shorter wavelength–higher frequency carriers are required to carry wideband signals, such as for television, long-distance communication for these wideband services is worse. Across terrestrial distances, a sequence of very expensive relay towers has been used. This was (and is) done to distribute television signals across the United States. However, nothing reasonable was able to solve the situation for communication across the oceans.

As soon as it became possible to place artificial Earth satellites into orbit, communication engineers began to develop communication satellites to solve this problem. In the early 1960s, there were a number of proposals as to how to best achieve this goal. *Passive* satellites, where ground transmission is just reflected off them, are very simple, but require large and powerful Earth stations. *Active* satellites receive the signal from Earth and reamplify it for transmission back to Earth. Although active satellites are more complicated electronically than passive satellites, they permit smaller and less powerful Earth stations. Currently all operational communication satellites are active.

Within the class of active satellites, the next consideration is the orbit. Low Earth Orbit (LEO) satellites generally can be lower powered and simpler than higher orbit satellites. Also, smaller boosters can be used to put them into orbit. Alternatively, multiple LEO satellites can be launched with a more powerful rocket. However, LEOs require a large number of satellites to achieve full Earth coverage and Earth station antennas must track each satellite across the sky and then switch to the next satellite coming over the horizon. Thus, LEOs have costly space segments and need Earth stations with costly antenna and control systems.

In October 1945, a brilliant science and science fiction writer, Arthur Clarke, wrote that there was a single, 24-hour-period orbit. A satellite in this orbit would appear stationary to an Earth observer—thus the term *geostationary*. This orbit at 22,300 miles above the Earth's surface was so high that three such satellites could cover almost the entire Earth. In the early 1960s, when many of the ideas about communication satellites were being developed, most of the scientists and engineers in the field believed that a geostationary satellite was many years in the future. SYNCOM 1 proved the naysayers wrong and today all operational communication satellites are geostationary.

This happened because a gifted trio of inventors (Dr. Harold Rosen, Thomas Hudspeth, and Donald Williams) at Hughes Aircraft Company conceived an elegantly simple spinning satellite that could be boosted to geostationary orbit by the rockets available at the time. Hughes funded the early developments, including a demonstration model of a geostationary satellite. After a considerable period of trying to interest customers in the project, Hughes was about to cancel the project. At that time, Donald Williams, one of the inventors of the concept, went to the office of Lawrence (Pat) Hyland, then the General Manager of Hughes Aircraft Company, and placed a cashier's check for $10,000 on Pat's desk. He told Mr. Hyland that the $10,000 was his life savings and that he wanted to contribute it to continuing the project. The Hughes Policy Board decided to continue the project and the rest is history.

Impressed by the vision of the Hughes prototype and encouraged by the U.S. Department of Defense (DoD), the National Aeronautics and Space Administration (NASA) awarded Hughes a contract to develop the SYNCOM series of satellites. The SYNCOM name is a derivation of "synchronous communication." And yes, Mr. Hyland did get Don Williams to take his check back.

In July 1963, SYNCOM 2, the first geosynchronous satellite, was successfully placed into geosynchronous orbit some 22,300 miles above the surface of the Earth, thus revolutionizing the field of communications. Later, its sister satellite, SYNCOM 3, became the first geostationary satellite (the distinction between geostationary and geosynchronous is explained in the next section of this chapter). SYNCOM 3 gained a great deal of attention by televising the Tokyo Olympics from Japan to the west coast of the United States.

Today, more than 35 years since the launch of SYNCOM 2, powerful satellites broadcast hundreds of digital television channels directly to the homes of consumers through an 18-inch receiver dish. These satellites are the direct descendants of the SYNCOM satellites.

Throughout most of the years from the first SYNCOM until the debut of the first Direct Broadcast Satellite (DBS) service, DIRECTV™ from Hughes Electronics, the signals going through the commercial communication satellites were analog. DBS brought the first wide-scale use of compressed digital audio visual services.

The DBS communication system takes advantage of parallel developments in a number of technologies. On the satellite side are more powerful transponders and shaped reflector antennas. On the content side, the Moving Pictures Experts Group (MPEG), a subunit of the International Standards Organization (ISO), created compression standards for audio and video compression, without which DBS would not have been technically or economically feasible.

In June 1994, DIRECTV launched a DBS service in the United States to provide customers with an alternative to traditional cable television networks, while at the same time eliminating the expense and inconvenience of traditional satellite television. In the 12 months that followed the initial DIRECTV subscription service, the satellite receiver became the number one best-selling consumer product of all time. In its first year, DIRECTV surpassed first-year VCR sales by a factor of 25! Within a year, over one million homes subscribed to DIRECTV's digital DBS service.

> *Personal Note: There at the Creation* Back in 1963, as a 26-year-old engineer working at Hughes Aircraft Company, I was assigned to design the Command Decoder System for the SYNCOM satellites and to be a part of the SYNCOM 1 launch team at Cape Canaveral. I was also fortunate to be involved in the development of DBS, particularly the MPEG standards.
>
> From July 1990 until August 1997, I was head of the Hughes delegation to MPEG and was able to help evolve the MPEG 1 and MPEG 2 standards so that they meet generic objectives and are directly applicable to DBS. Thus, I have had the incredible good fortune to be there at the creation of both the geostationary satellite business and its progeny—DBS.

1.2 Arthur C. Clarke's Vision

It is instructive to review how Arthur C. Clarke arrived at his geostationary concept—EXTRA-TERRESTRIAL RELAYS. Astronomers have studied the heavens for thousands of years. It wasn't until 1609 to 1611, however, that the German astronomer Johannes Kepler published his three laws of orbital motion. These laws can be used to calculate the altitude and orbital period for any satellite orbit, whether it is that of the Earth around the sun, or a DBS around the Earth.

Kepler's laws can be used to show the period of a satellite orbiting a body:

$$T = \frac{2\pi}{\sqrt{M}} a^{3/2} \tag{1.1}$$

where a is the semimajor axis (the radius for a circular orbit) and M is the gravitational constant. For Earth,

$$\begin{aligned} M &= 1.408 * 10^{16}\ \mathrm{ft^3/sec^2} \\ &= 3.987 * 10^{13}\ \mathrm{m^3/sec^2} \end{aligned} \tag{1.2}$$

Because the only two variables in (1.1) are a and T, every orbital period has a unique orbital height. Thus, there is one orbital height in which the orbital period is 24 hours. In this orbit, a satellite at this altitude will appear stationary to an observer on Earth. As a result, this orbit is called geostationary. See the following discussion for the distinction between geosynchronous and geostationary satellites.

Solving (1.1) for a yields

$$\begin{aligned} a^{3/2} &= \frac{TM^{1/2}}{2\pi} \\ a &= \left(\frac{T}{2\pi}\right)^{2/3} M^{1/3} \end{aligned} \tag{1.3}$$

For a geostationary orbit

$$T = 24\ \mathrm{h} = 8.64 * 10^4\ \mathrm{sec}$$

Inserting the values for T and M into (1.3) yields

$$a = 13.86 * 10^5\ \mathrm{ft, or}$$
$$a = 26{,}250\ \mathrm{miles}$$

from the center of the Earth. Since the radius of the Earth is approximately 4,000 miles, the orbital height above the surface of the Earth is 22,250 miles.

Arthur C. Clarke is certainly the most gifted science-fiction writer of the twentieth century. *2001: A Space Odyssey* is perhaps his most famous work, although he wrote many more books. To describe Clarke as a science-fiction writer is misleading, however. While most science fiction describes things that are physically impossible or will never happen, Clarke writes science stories that predict the future. Thus, "visionary futurist" would be a more fitting description of this man.

In 1945, Clarke wrote an article in *Wireless World* magazine that pointed out that three communication satellites (Clarke called them EXTRA-TERRESTRIAL RELAYS) in geostationary orbit and separated by 120° longitude could provide ubiquitous communications to almost every place in the world [Clarke45]. Clarke later wistfully noted in his book *Voices from the Sky* that he had mixed feelings about originating one of the most commercially viable ideas of the twentieth century and selling it for just $40 [Clarke65].

Clarke never patented his idea because he felt it would be well into the next century before these satellites could be built and the technology realized. In reality, it took less than 20 years to build the prototype. Clarke has said that if he had not conceived and published his ideas about the geostationary satellite, someone else would have within ten years. Although such projections are always speculative, I believe the combined genius of Clarke and Dr. Harold A. Rosen, who led the group that developed the first geostationary satellite, advanced the realization of the first geostationary satellite by at least several decades. Historic justice has been served in that these two giants became friends in their later years and Dr. Rosen won the Arthur C. Clarke Award for contribution to space science three times.

Sometimes the words *geosynchronous* and *geostationary* are used synonymously, but this is incorrect. If a satellite is in a 24-hour orbit, it is said to be geosynchronous; however, if the orbital plane and the equatorial plane of the earth are not the same, the satellite will not appear stationary to an observer on Earth. A satellite so placed traverses a trajectory on the Earth in the shape of a figure eight every 24 hours. Thus, it is exactly on the equator at only two points in time each day. If, however, the plane of the orbit and the equatorial plane are identical, the satellite is said to be geostationary, and, indeed, it appears stationary to an observer on Earth. It is important to have the DBS satellites geostationary so that no antenna movement is required to keep the satellite within the Earth station beam.

1.3 The Early Days of Satellite Communications

The rockets developed by Germany during World War II were not powerful enough to launch an object into orbit around the Earth, much less into geostationary orbit, but it was clear that improvements in these rockets would enable them to do so. This was the reason there was such competition between the USSR and the United States to "obtain" the German rocket scientists at the end of the war. It also was clear that an artificial Earth satellite could be used as EXTRA-TERRESTRIAL RELAYS, to use Clarke's term. The ability to orbit such satellites became evident when the former USSR shocked the world by orbiting the Sputnik satellite in 1957.

During the period from 1945 to 1957 and beyond, the United States, with Werner von Braun and other expatriate German rocket scientists, was working desperately to develop satellites and the rockets to launch them. The USSR was the first to orbit a man-made satellite and to put a man in orbit, however. In those early days there were several competing ideas of what a communication satellite should be. One of the first considerations was it should be passive or active.

The idea behind the passive satellite was to orbit a relatively large reflector that could be used to reflect the ground-transmitted signal back to the Earth. The passive reflector satellite requires a minimum of electronics but it creates a number of problems. Perhaps the most serious is that the signal needs to travel twice as far without amplification. The power density of a radio frequency signal decreases by the square of its distance from the emitter. Thus, doubling the distance reduces the power at the Earth receiver by a factor of four compared with what could be received by an antenna on the satellite. While the Echo satellite program did launch several aluminum-covered balloon satellites, the program did not go further because of the very large, complicated, and powerful ground stations required.

An alternative to passive satellites was active satellites. Active satellites receive a signal from an Earth station, translate the frequency of the signal, amplify it, and retransmit it to Earth. Active communication satellites can be placed in a variety of orbits. New orbital concepts are still being conceived and tried today. The first orbit that was tried for active communication satellites was what is now called a low Earth orbit, or LEO. Referring to Equation (1.1), the lower the orbital altitude, the shorter the orbital period. While the term LEO can cover a variety of orbital altitudes, assume a 100-mile orbital altitude that corresponds to a 90-minute orbital period. Since the satellite traverses 360 degrees in 90 minutes, the rate is four degrees per minute. Thus, it moves fairly rapidly across the sky.

For example, assume that the satellite is mutually visible for an arc of 60 degrees. The duration of this visibility will be only 15 minutes. Note that both the transmitter and the receiver Earth stations must have movable antennas that track the satellite across the sky. Also, to have continuous coverage between stations, a large number of LEO satellites would be required and a complex control scheme would be needed to hand off the transmission from one satellite disappearing over the horizon to the next one coming into view.

In July 1962, AT&T orbited the Telstar satellite, the world's first active communication satellite. Telstar worked well and proved a number of theories. However, as an LEO, the number of satellites that would be required for a complete-coverage network and the complexity of the ground antennas that were required worked against the Telstar concept, not allowing it to progress very far.

It was generally acknowledged that if Clarke's vision of geostationary satellites could be realized, it would be the ideal solution. However, virtually all of the scientists and engineers of the time believed that a geostationary system was so complicated it would take years before one could be orbited. That was not to be the case. Led by Rosen, a small group of engineers at Hughes Aircraft Company developed a prototype geostationary satellite. Impressed by the vision of this prototype and encouraged by the DoD, the National Aeronautics and Space Administration (NASA) awarded Hughes a contract to develop the SYNCOM series of satellites.

1.4 SYNCOM

As noted, the engineers and scientists involved in the early development of communication satellites were almost unanimous in their belief that geosynchronous/geostationary satellites were far in the future. What was it about SYNCOM that made them all wrong? The following sections give a general overview of the SYNCOM satellites.

1.4.1 Stabilization

The SYNCOM satellites were the shape of a cylinder and were stabilized by spinning the satellite about the cylindrical axis. If the satellite had an orientation in inertial space, conservation of angular momentum required that it retain that orientation unless acted upon by a force.

1.4.2 Attitude Control and Orbital Positioning

For a satellite in the form of a spinning cylinder, all of the attitude control and orbital positioning can be achieved by two thrusters:

Thruster 1 is parallel to the spin axis but offset from it

Thruster 2 is orthogonal to the spin axis and thrusting through the satellite center of gravity

This form of control was invented by Don Williams and is the basis for the famous "Williams patents."

On the SYNCOM satellites, the thrusters were provided by nitrogen compressed to very high pressure and stored in small tanks. Small jets released the nitrogen to provide the thrust.

1.4.3 Power

Power for the satellites was provided by solar cells that covered the outside of the satellite. A bluish color comes from the satellite's solar cells. Batteries provided the power when the satellites suffered from solar eclipses.

An added benefit of the spin stabilization was the fact that the solar cells could cool off when they were on the side of the satellite away from the sun. Solar cell efficiency was high because the solar cells never became too hot.

1.4.4 Orbital Injection

The three-stage Delta booster could launch the SYNCOM satellites into a highly elliptical orbit. When the satellite reached its apogee of this elliptical orbit (22,300 miles) an integral fourth stage was fired, circularizing the orbit.

1.4.5 Communication Antenna

The communication antenna was a simple dipole located coaxially with the satellite spin axis. The radiation pattern, orthogonal to the spin axis, is in the shape of a doughnut. Since the Earth subtends an angle of about 18 degrees from geostationary orbit, only 18/360, or 5 percent, of the radiated power hits the Earth.

Although inefficient, the dipole was simplicity itself. The whole story of the evolution of satellite antennas is one of relentless progress (see Chapter 4). Figure 1.1 is a photograph of an actual SYNCOM satellite.

SYNCOM 1 was launched February 14, 1963, and was lost five hours after launch when its apogee motor firing caused a complete failure. The exact cause of the SYNCOM 1 failure probably will never be known. Certainly it was an on-board explosion: either the integral fourth stage exploded or one of the nitrogen tanks was scratched, turning it into a lethal projectile.

Optical telescope images seemed to indicate the satellite was in a higher orbit than was planned. This would mean the satellite lost mass during fourth stage burn, which would point to the nitrogen tanks as the cause of the explosion.

Regardless, the SYNCOM 1 experience showed there was nothing fundamentally wrong with the SYNCOM design. After the initial failure, certain engineering changes were made to the next flight satellite but nothing fundamental was changed. SYNCOM 2 was launched on July 26, 1963, and became the first successful geosynchronous satellite.

It probably is difficult for most people, who now take for granted having satellite TV beamed into their homes, to appreciate the significance of this event. It changed the world forever. A strong argument can be made that it

Figure 1.1 Photograph of a SYNCOM Satellite Source: Reproduced from *Vectors,* XXX(3): Cover, 1988. Copyright © 1988 Hughes Aircraft Co. All rights reserved. Used with permission.

ultimately helped lead to the defeat of Communism. Try as they might, the Communist rulers could not stop their citizens from receiving satellite TV. When people in Communist countries saw what life was like in other parts of the world, they demanded the same freedoms.

In Section 1.1, I explained the unreliability of long-distance communication. The dramatic and immediate improvement permitted by geosynchronous satellite communication is illustrated by an event that took place shortly after the SYNCOM 2 launch. In 1963, the *USS Kingsport,* stationed in Lagos Harbor, Nigeria, conversed with the U.S. Navy base at Lakehurst, New Jersey. It was the first voice message ever transmitted between continents by a synchronous satellite. Later came the highly publicized telephone conversation between U.S. President John F. Kennedy and Nigerian Prime Minister Abubakar Balewa. It should be noted that at that time, all the normal voice circuits were down because of atmospheric conditions but the satellite circuit worked perfectly! [Roddy96]

SYNCOM 2 later was repositioned over the Pacific, where it provided extensive communications capability during the war in Vietnam.

SYNCOM 3, the last of the series, was launched on August 19, 1964. By using a more powerful fourth stage, it was possible to make the satellite's orbital plane the same as the Earth's equatorial plane. Thus, SYNCOM 3 became the world's first geostationary satellite. SYNCOM 3 later gained fame by telecasting the Tokyo Olympics across the Pacific. Figure 1.2 shows the

Figure 1.2 Antenna Used to Receive the 1964 Olympic Games from Tokyo
Source: Reproduced from *Vectors*, XXX(3): 3, 1988. Copyright © 1988 Hughes Aircraft Co. All rights reserved. Used with permission.

85-foot-diameter antenna used to receive the signal at Point Mugu, California. Note the man on the pedestal next to the antenna. Today's DBS antenna is smaller in diameter than the length of his arm.

Once the geostationary satellite concept had been proven, it was clear that this would be the preferred approach. After 30 years, this concept continues to dominate the communication satellite business, particularly for broadcast purposes.

1.5 The Early Commercial Geostationary Earth Orbit Satellites: INTELSAT I and II

In the early 1960s it became apparent that communication satellites would benefit not only the citizens of the United States, but of the world as well.

In 1963 the U.S. Congress formed the Communication Satellite Corporation (COMSAT) as a quasiprivate company to represent the interests of the United States in international communication satellite matters. Dr. Joseph Charyk was appointed the founding president of COMSAT and served with distinction as president, CEO, chairman of the board, and director for 35 years.

Founded in 1964, the International Satellite Consortium (INTELSAT) was the first organization to provide global satellite coverage and connectivity, and continues to be the communications provider with the broadest reach and the most comprehensive range of service.

INTELSAT is an international, not-for-profit cooperative of more than 140 member nations. Owners contribute capital in proportion to their relative use of the system and receive a return on their investment. Users pay a charge for all INTELSAT services, depending on the type, amount, and duration of the service. Any nation may use the INTELSAT system, whether or not it is a member. Most of the decisions that member nations must make regarding the INTELSAT system are accomplished by consensus—a noteworthy achievement for an international organization with such a large and diverse membership. For more information, the reader is referred to INTELSAT's Web page.

It was important that the fledgling INTELSAT establish a real system as early as possible. COMSAT and INTELSAT made the decision that their first system would be geostationary and contracted with Hughes to build the first commercial communication satellite.

On the twenty-fifth anniversary of the first commercial communication satellite, Charyk noted that he made the historic decision to use geostationary Earth orbit satellites (GEOS) because of Rosen and the Hughes team. This first commercial communication satellite, known officially as INTELSAT 1 and affectionately called Early Bird, was launched on April 6, 1965, and provided either 2,400 voice circuits or one two-way television channel between the United States and Europe. During the 1960s and 1970s, message capacity and transmission power of the INTELSAT 2 (Blue Bird), 3, and 4 generations were progressively increased by beaming the satellite power only to Earth and segmenting the broadcast spectrum into transponder units of a certain bandwidth.

1.6 The Evolution of Communication Satellites

In the 33 years since the launch of SYNCOM 1 and 2, rocket science has continued to develop and provide for larger and larger payloads. Perhaps nothing illustrates this more dramatically than Table 1.1, which shows the evolution of satellites from SYNCOM to date. Every subsystem on the satellite has evolved remarkably from where it was 33 years ago. Improvements have been made to the solar cells, batteries, propellants for station keeping, and all of the electronics technologies. The last line of Table 1.1 shows the figures for the HS 702 satellites, which are expected to be placed in service by the end of the decade.

As can be seen from Table 1.1, a new era in satellite size and power began in 1988 with the launch of the OPTUS-B satellite. With this satellite, weight

Table 1.1 The Evolution of Communication Satellites

Year	Satellite	Weight (lbs)	Power (watts)
1963	SYNCOM	78	19
1966	INTELSAT II	169	70
1967	ATS 1	775	120
1969	INTELSAT IV	1,610	425
1973	COMSTAR	1,746	610
1981	PALAPA-B	1,437	1,062
1987	BSB	1,450	1,000
1988	OPTUS-B	2,470	2,915
1993	DBS-1	3,800	4,300
1999	HS 702*	payload = 2,200	up to 15,000

*Data is not available to relate the HS 702 to previous models.

was increased 1,000 pounds and power was almost tripled. The OPTUS-B and DBS satellites are Hughes Electronics's HS 601 body stabilized models. As shown in Figure 1.3, the solar arrays are 86 feet (26 meters) from tip to tip and generate 4,300 watts of prime power. The solar arrays are connected to a 32-cell nickel-hydrogen battery for uninterrupted service during eclipses.

Body Stabilized: The spin-stabilized satellites permitted the early development of geostationary communication satellites. However, the basic solar-cell power of a spinning satellite is controlled by the surface area of the cylinder. Satellite development reached the point where this area was not large enough for the demands of more powerful satellites.

The solution has been the body-stabilized satellite. The solar cells are located on large paddles that can be continuously adjusted to maximize the power output from the sun (see Figure 1.3). Attitude stabilization is achieved by a flywheel in the body of the satellite. Thus, in some sense, spin stabilization lives on.

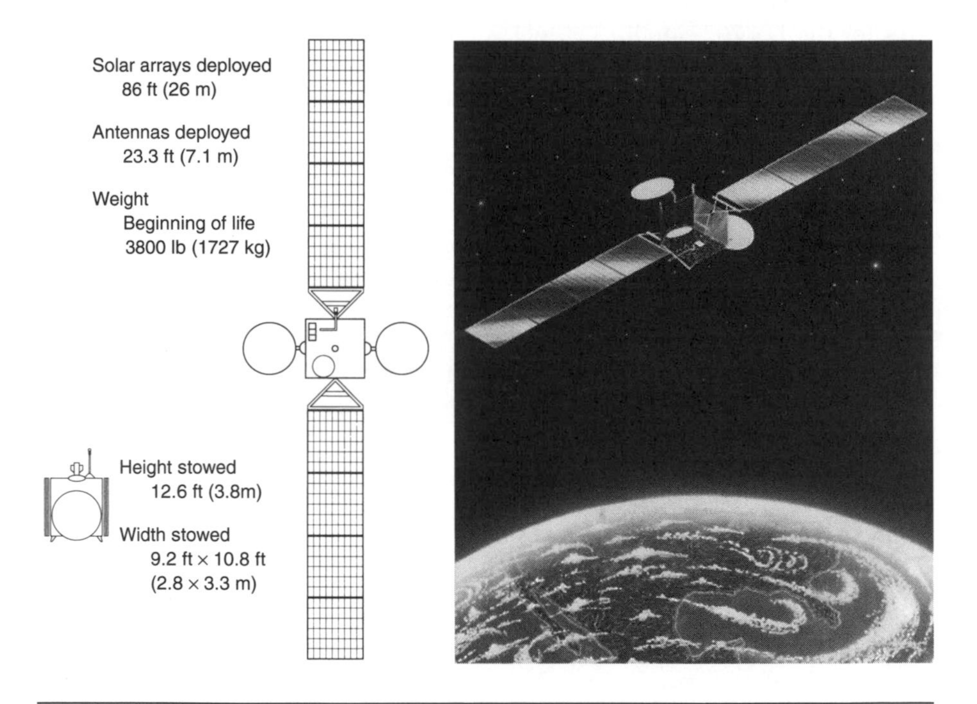

Figure 1.3 The Hughes DBS Satellites Source: Reproduced from Hughes Space & Communications DBS brochure (HSC95D104/1500/4-95), 1995. Used with permission.

1.7 The Hughes Direct Broadcast Satellites

On December 17, 1993, an Ariane 4 rocket launched DBS-1 from Kourou, French Guiana, ushering in a new era in satellite communications. On August 3, 1994, DBS-2 was launched, and on June 10, 1995, DBS-3 was launched. All are located at 101° W longitude.

Figure 1.4 shows the DBS-1 satellite in its stowed position (solar panels and antennas folded in), and Figure 1.5 shows the actual liftoff. The HS 601 body is composed of two main modules: the bus module and the payload module. The bus module is the primary structure that carries launch vehicle loads and also carries the propulsion, attitude control, and electrical power subsystems. The payload module is a honeycomb structure that contains the payload electronics, telemetry, command, and ranging subsystems. The HS 601 transmit-and-receive antenna reflectors are eight feet in diameter. They are constructed with a composite graphite material and weigh only 20 pounds each. Both are fed from their own feedhorn. For more on this innovative antenna design, see section 4.2 in Chapter 4.

Figure 1.4 DBS-1 in Stowed Position Source: Reproduced from Hughes Space & Communications DBS brochure (HSC95D104/1500/4-95), 1995. Used with permission.

Figure 1.5 DBS Liftoff Source: Reproduced from Hughes Space & Communications DBS brochure (HSC95D104/1500/4-95), 1995. Used with permission.

1.8 Frequency Bands

The frequency bands for communication satellites certainly have some of the most confusing nomenclature in modern science. The bands are designated by alpha characters, which have their origins in World War II secrecy considerations and have become so embedded in the language that they are still used. Table 1.2 shows band designations and their corresponding frequency range.

Most of the early communication satellites were C-Band, with an uplink at 6 GHz and a downlink at 4 GHz. Even today the mainstay of communication satellite distribution of television and the corresponding backyard dish products are in the C-Band. C-Band has a number of highly desirable attributes, including the fact that it has very little attenuation because of rain and

Table 1.2 Band Designation

Band	Frequency
L-Band	1–2 GHz
S-Band	2–4 GHz
C-Band	4–8 GHz
X-Band	8–12 GHz
K_u-Band	12–18 GHz
K_a-Band	18–30 GHz

the componentry is inexpensive. However, the world is running out of space on the C-Band spectrum. The satellites in geosynchronous orbit can be placed only so close to each other without causing undesirable interference. Thus, many of the new services, including DBS, are provided in the K_u-Band. Most of this book will discuss the communication system in K_u-Band; however, Chapter 12 discusses the move to K_a-Band.

1.9 Summary

A large number of technological capabilities came together at the right time to make DBS economically viable. Certainly the satellite size and prime power-enabling high-power transponders were essential, and the creation of the MPEG 1 and MPEG 2 compression standards could not have come at a better time.

Gordon Moore, chairman of Intel, made the prediction more than 20 years ago that every two years the cost of semiconductors for the same function would halve; that is, one would get twice the functionality for the same price. This certainly must rank as one of the most remarkable predictions of all time. Much of our technology today is made possible by affordable integrated circuits. The inexorable march of Moore's law made the cost of the DBS settop unit affordable.

A number of other design decisions vital to the success of DBS had to be made, however. These are explored in Chapters 3 through 9.[Clarke65] Clarke, Arthur C., *Voices From the Sky*, New York: Harper & Row, 1965/

References

[Clarke45] Clarke, Arthur C., "EXTRA-TERRESTRIAL RELAYS," *Wireless World*, 51(10):305–308, October 1945.

[Clarke65] Clarke, Arthur C., *Voices From the Sky*, New York: Harper & Row, 1965.

[Hughes88] Hughes Electronics, *Vectors*, XXX(3): Cover, 2–8, Hughes Aircraft Company, Torrance, CA, 1993.

[Hughes95] Hughes Space & Communications, DBS Brochure (HSC950104/1500/4-95), Hughes Aircraft Company, April 1995.

[Roddy95] Roddy, Dennis, *Satellite Communications*, 2nd ed., New York: McGraw-Hill, 1996.

CHAPTER TWO

Regulatory Framework

This chapter explains where a number of the fixed attributes of DBS come from, including uplink frequencies, downlink frequencies, polarization, and orbital positions. The rest of this book discusses what has been done to maximize the services that can be provided by DBS.

The orbital locations, spectrum, and other parameters were established by international agreement. The Federal Communications Commission (FCC) then granted licenses to companies for transponders at the orbital locations granted to the United States by the international agreements.

The original licensees received the transponder licenses at no cost (other than legal fees). The most recent reallocation of these transponders was done by auction. The results are described in section 2.3.

2.1 The 1977 World Administrative Radio Council and 1983 Regional Administrative Radio Council

In 1977, a World Administrative Radio Council (WARC) made the farsighted decision to allocate spectrum and set parameters for a DBS service also called Broadcast Satellite Service (BSS). The 1977 WARC made the allocations for region 1 (Europe) and region 3 (Asia and Oceana), and made a preliminary plan for region 2, the Western Hemisphere [WARC77].

One of the critical decisions was that the orbital spacing for clusters of satellites would be 9 degrees. This has far-reaching implications for overall DBS capacity.

In 1983, a Regional Administrative Radio Council (RARC) made the comparable decision for region 2, which includes the United States and the rest of the Western Hemisphere. Eight orbital positions were set aside for the United States. The spectrum was established as 17.3–17.8 GHz for the uplink and 12.2–12.7 GHz for the downlink. Thirty-two frequencies were allocated at each orbital slot [RARC83].

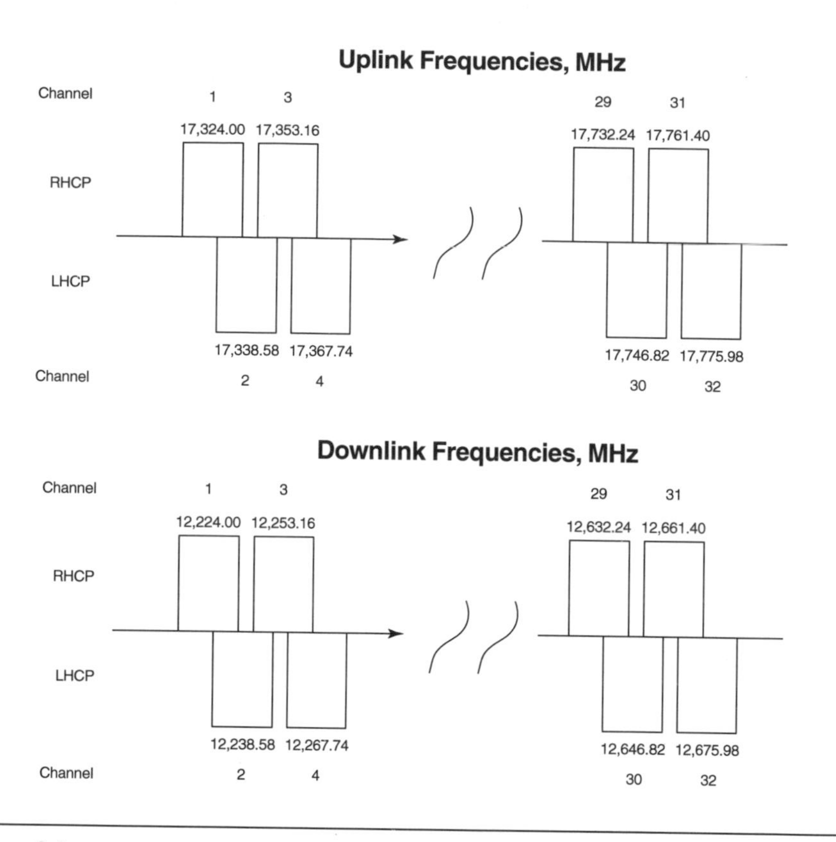

Figure 2.1 Region 2 DBS Frequency Plan Source: Adapted from Proceedings published by the Regional Administrative Radio Council in 1983.

The RARC also made a provision for frequency reuse. Figure 2.1 shows the complete frequency plan. As can be seen, there are 16 transponders with right-hand circular polarization and 16 transponders with left-hand circular polarization. This permits 32 24-MHz bandwidth transponders in the 500-MHz DBS band with 5-MHz spacing between transponders. The Effective Isotropic Radiated Power (EIRP) was established as 57 dBW.

2.2 Federal Communications Commission Licensing

The following orbital locations are the ones allocated to the United States for which the FCC granted licenses for transponders: 61.5° West, 101° West, 110° West, 119° West, 148° West, 157° West, 166° West, and 175° West.

Initially, the first four orbital locations in this list were called the Eastern locations and the last four were called the Western locations, because it was believed the satellites would not be powerful enough to cover the entire Continental United States (CONUS). However, DIRECTV™ and United States Satellite Broadcasting (USSB) proved that satellites could be powerful enough to cover CONUS. Subsequently, the orbital locations at 101°, 110°, and 119° W longitude were called "full CONUS" and the others "partial CONUS."

Note that the orbital location at 61.5° W longitude was designated full CONUS at one time because it is possible for U.S. West Coast–based receivers to see a Geostationary Earth Orbit (GEO) at 61.5° W longitude. The elevation angles are so small, however, that a high percentage of locations are blocked by trees, buildings, hills, and so forth. So the FCC relented and agreed to call the 61.5° W longitude slot a partial CONUS.

2.3 Recovery and Reallocation of Spectrum

Advanced Communications Corporation (ACC) received one of the conditional construction permits issued in 1984 for 27 full-CONUS transponders at 110° W longitude and 24 partial-CONUS transponders at 148° W longitude.

In the fall of 1995 the FCC determined that ACC had not exercised due diligence in moving towards a DBS system and reclaimed all 51 transponders. The FCC then decided that the most equitable way to distribute these transponders was by auction.

It also should be noted that the FCC never allocated one of the transponders at 110° W longitude. The FCC added this transponder to the 27 recovered from ACC to create a package for the auction.

The 28 full-CONUS transponders at 110° W were auctioned off in the first quarter of 1996. The winning bid was an astounding $682 million dollars by MCI. Ironically, during the FCC's preparation for the auction, one of the FCC commissioners warned his colleagues not to be swayed by the fact that one of the prospective bidders was prepared to bid the unbelievable sum of $175 million. The winning bid was almost four times this.

Echostar subsequently won an auction for the 24 transponders at 148° W longitude partial-CONUS orbital slots.

Figure 2.2 and Table 2.1 show the licensees at each of the U.S. orbital locations and the percentage of DBS transponder ownership by these licensees.

Partial CONUS				Full CONUS			Partial CONUS
175 WI	**166 WI**	**157 WI**	**148 WI**	**119 WI**	**110 WI**	**101 WI**	**61.5 WI**
Echostar/ Directsat 11 Channels	Echostar/ Directsat 11 Channels	DirecTV 27 Channels	Echostar 24 Channels	Echostar/ Directsat 21 Channels	MCI 28 Channels	DirecTV 27 Channels	Continental* 11 Channels
DBSC** 11 Channels	Continental 11 Channels		USSB 8 Channels	TCI/Tempo 11 Channels	USSB 3 Channels	USSB 5 Channels	DBSC** 11 Channels
	Dominion 8 Channels				Echostar/ Directsat 1 Channel		Dominion 8 Channels
Unassigned 10 Channels	Unassigned 2 Channels	Unassigned 5 Channels					Unassigned 2 Channels

*Loral holds a controlling interest
**DBSC of Delaware

Figure 2.2 Current Allocations of U.S. DBS Channels Source: From the Consumer Project on Technology, Center for the Study of Responsive Law, Washington, DC, March 15, 1996 (see also *http://www.essential.org/cpt/telecom/echo.txt*).

Table 2.1 U.S. DBS Channel Ownership

	Full CONUS		Partial CONUS		Total	
MCI	28	29%	0	0%	28	11%
DIRECTV	27	28%	27	17%	54	21%
Echostar	22	23%	46	29%	68	27%
TCI/Tempo	11	11%	11	7%	22	9%
USSB	8	8%	8	5%	16	6%
Continental*			22	14%	22	9%
DBSC**			22	14%	22	9%
Dominion			16	10%	16	6%
Unassigned			8	5%	8	3%
Echostar/DBSC (if merger is approved)	22	23%	68	43%	90	35%

*Loral holds a controlling interest
**DBSC of Delaware
Source: From the Consumer Project on Technology, Center for the Study of Responsive Law, Washington, DC, March 15, 1996 (see also *http://www.essential.org/cpt/telecom/echo.txt*).

References

[FCC95a] FCC Notice of Proposed Rulemaking, "In the Matter of Revision of Rules and Policies for the Direct Broadcast Satellite Service" (FCC 95-443), October 27, 1995.

[FCC95b] FCC Report and Order, "In the Matter of Revision of Rules and Policies for the Direct Broadcast Satellite Service" (FCC 95-507), December 15, 1995.

[IPN96] Information Policy Note, from an Internet newsletter (available from *www. listproc@tap.org*), April 2, 1996.

[RARC83] RARC, Proceedings of the 1983 Regional Administrative Radio Council, Geneva, Switzerland.

[WARC77] WARC, Proceedings of the 1977 World Administrative Radio Council, Geneva, Switzerland.

CHAPTER THREE

An Overview of the DBS Communication System

3.1 Overview

Figure 3.1 is a simplified schematic/block diagram of the DBS communication system. There is a separate identical system for each transponder. Starting at the left of Figure 3.1, source materials from a variety of sources are delivered to the uplink facility, including analog satellite backhaul, cable, fiber, and packaged goods (analog or digital).

The incoming source materials are input to video and audio compressors. The outputs of the video and audio compressors are Packetized Elementary Streams (PES), which are input to the multiplexer. A number of separate services are Time Division Multiplexed (TDM) into a single bitstream for each transponder. Chapter 7 discusses the formation, multiplexing, and transport of PESs in detail.

As shown in Figure 3.1, other data can be multiplexed into the bitstream for each transponder. Program Guide information and Conditional Access entitlement messages are the most important sources in this category, but other functions, such as e-mail messages, also can be included.

The Forward Error Correction (FEC) then is applied to the bitstream from the multiplexer. As discussed in detail in Chapter 5, a Reed-Solomon block code is applied first, followed by interleaving and a convolutional code.

The final bitstream (information bits and FEC parity bits, sometimes called the chip stream) is input to the Quaternary (Quadrature) Phase Shift Keying (QPSK) modulator. In a digital communication system, digital modulation usually is employed, which means one of a finite set of waveforms is transmitted in a basic interval called the symbol time. In QPSK there are four possible waveforms, each of which represents two bits. This method is discussed in more detail in Chapter 4 and Appendix B.

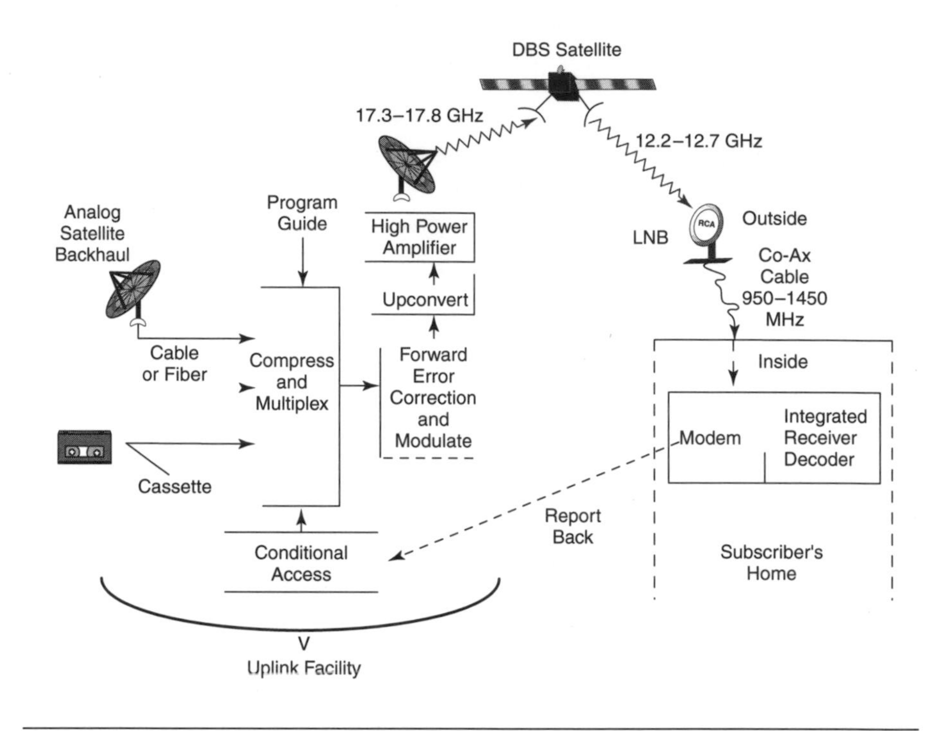

Figure 3.1 Overall DBS Communication System (for each transponder)

The chip stream then is converted to the appropriate frequency for that transponder (between 17.3 and 17.8 GHz), amplified, and transmitted to a DBS satellite.

The DBS satellite "transponds" the signal and retransmits it to earth. Figure 3.2 shows that the transponder function consists of converting the uplink frequency to an output, or downlink, frequency (it also provides amplification) that is 5.1 GHz lower. Because the role of the DBS satellite is to translate the signal to a different frequency and then amplify it, the satellite is frequently referred to as a "bent pipe." It should be noted that, consequently, the signal is not regenerated in the satellite. Conceptually, this transponding can be performed by the heterodyne circuit shown in Figure 3.2.

The 12.2–12.7-GHz signal is received via an 18-inch reflector antenna and offset Low-Noise Block (LNB). The LNB translates the received signal to L-Band (950–1,450 MHz) and amplifies it to drive the coaxial cable to the Integrated Receiver Decoder (IRD). More than 500 feet of coaxial cable can be run from the LNB to the IRD without further amplification.

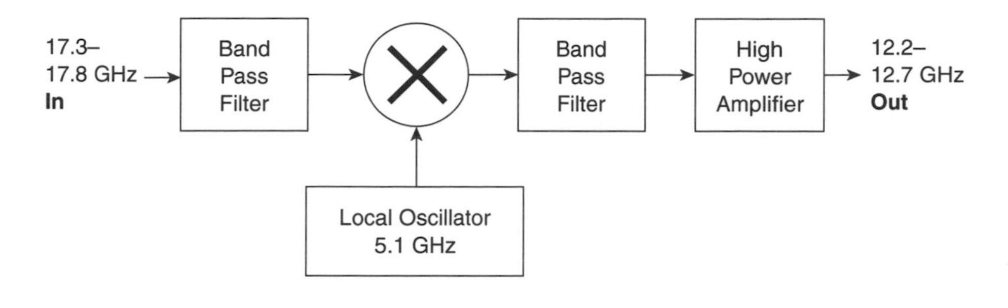

Figure 3.2 Satellite Transponder Functional Diagram

The IRD (see Chapter 11) basically reverses the uplink process and consists of the following:

Tuner—selects one of the 32 transponders

Demodulator—selects which of the four possible subcarrier waveforms was transmitted, putting out two bits per symbol

Forward error correction decode—removes the concatenated FEC

Demultiplexer—extracts the selected service from the overall TDM bitstream

Buffer—small storage of demultiplexed bytes between demultiplexer and decompression engines

Decompression engines—decompressors for audio and video

A key feature of Hughes Electronics' Digital Satellite System (DSS™) is that the uplink can individually request an IRD to initiate a telephone call back to the management center. This feature allows the viewing-history storage to be transmitted to the management center for billing purposes. Note that this allows true impulse pay-per-view operation. The subscriber just selects the service to watch. The viewing-history memory remembers this information and sends it to the management center when it is required.

3.2 Multiple Access and Multiplexing

Most studies have shown that to be economical, a DBS service in the United States must offer between 150 and 200 independent A/V channels. As pointed out in Chapter 1, each orbital slot for DBS has 32 transponders, each with 24 MHz of bandwidth; thus, between five and seven channels have to be allocated per transponder.

Two techniques exist for performing this allocation:

Frequency division multiplexing (FDM)—Each of the independent channels to be shared within a transponder can be assigned a specific frequency within the bandwidth of that transponder. Each of the services then shares the carrier power of that particular transponder.

Time division multiplexing (TDM)—An alternative to FDM is to have time slots for each of the services and to separate the services by allocating different time slots to them.

Each technique has advantages and disadvantages. FDM has the advantage that multiple uplinks can be employed with minimal coordination. However, FDM suffers from the basic physics of the satellite transmitter. The high-power amplifier has either a traveling wave tube amplifier (TWT-A) or a solid-state amplifier, both having the so-called s-curve when power out versus power in is plotted.

As shown in Figure 3.3, the sum of the different carriers in an FDM system must operate within the near-linear region of the s-curve. Thus, the FDM system has to have what is called backoff, whereby the power is backed off so that the operation is assured in the linear portion of the s-curve. Typical backoffs as high as 10 dB have been reported; thus, FDM suffers a major drawback compared to TDM.

On the other hand, TDM requires that virtually all of the source signals be uplinked from a single location (at least for each transponder). The problem with multiple uplink sites in TDM is the fact that each uplink would have to time its transmission so that it arrived at the satellite at exactly the right time slot. While this can be done with appropriate use of guard bands and beacons back from the satellite, it creates a less-efficient use of the available bandwidth and potentially complicates the receiver, neither of which can be tolerated in a DBS system.

For these reasons, all of the known DBS systems employ time division multiplexing with a single uplink facility. While the uplinks could be separated by transponder with no negative effects, the DBS providers elected to have all of their transponder uplinks at the same location because of the infrastructure required to support the uplinking process. Chapter 10 describes a typical uplink facility.

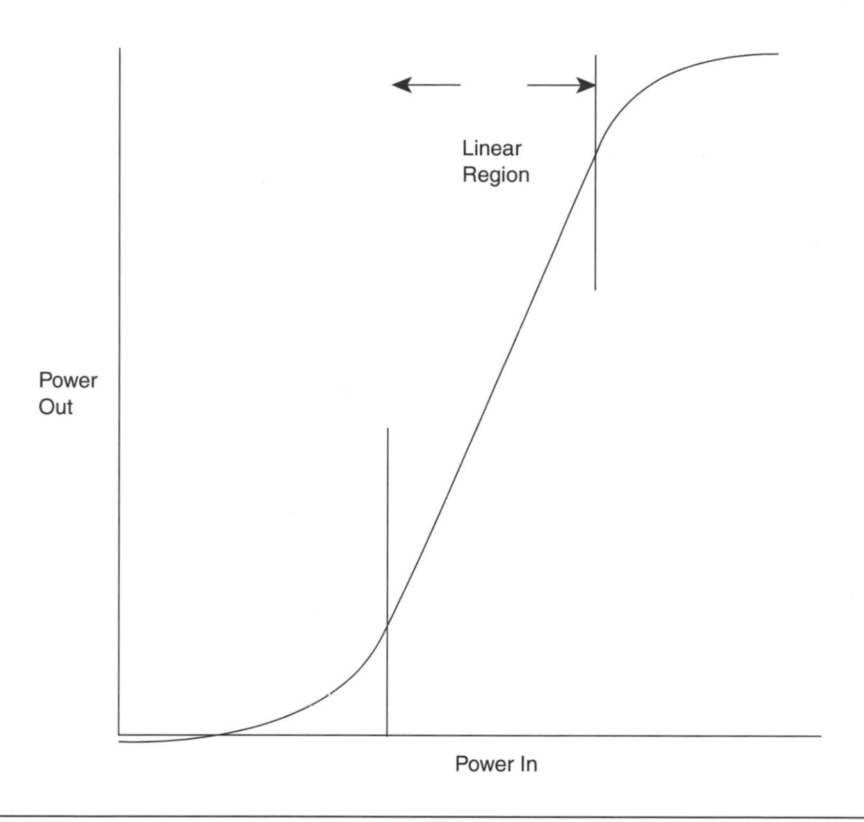

Figure 3.3 S-Curve for High-power Amplifier

3.3 Transponder Power and Configuration

The Hughes DBS satellites can be configured as either 16 transponders at 120 watts each (16/120) TWT amplifiers (TWT-A) or as 8 transponders at 240 watts each (8/240). With the three-DBS-satellite configuration of the Hughes DBS satellites at 101° W longitude, the odd transponders are 240 watts per 6/7 rate convolutional code and the even transponders are 120 watts per 2/3 rate convolutional code.

3.4 The Throughput

As covered in Chapter 2, the analog transponder bandwidth is 24 MHz. Section 4.3 of Chapter 4 will show that QPSK is used for all known DBS systems, which yields 2 bits per symbol.

Experience has shown that 20 megasymbols per second can be transmitted through the 24-MHz analog bandwidth. Thus, we have

$$20 \text{ megasymbols per sec} * 2 \text{ bits/symbol} = 40 \text{ megabits/sec}$$

In the case of DIRECTV, the Reed-Solomon outer code is (146, 130),[1] and there is one sync byte. The effective code rate for the outer code is thus 130/147. This Reed-Solomon code can correct up to 8 byte errors in each 147-byte packet.

DVB utilizes the MPEG 2 Systems packet structure, which is 188 bytes in length. It then utilizes a shortened (204,188) Reed-Solomon code. The inner code can be selected from five different values: 1/2, 2/3, 3/4, 5/6, and 7/8. The overall code rate can be selected from 188/204 times any of the five inner-code rates.

For DIRECTV the inner convolutional code can take on two values: 2/3 or 6/7, depending on the transponder power. Thus, the overall code rate is

$$\frac{130}{147} * \frac{2}{3} = .5896, \text{ for the low-power case}$$
$$\frac{130}{147} * \frac{6}{7} = .758, \text{ for the high-power case}$$

The corresponding information throughputs are 40 Mbps * 0.5896 = 23.58 Mbps for the low-power case, or 40 Mbps * 0.758 = 30.3 Mbps for the high-power case.

3.5 Overall Throughput

As of this writing, there are three DBS satellites at 101° W longitude. Two are configured as 8/240 and one is configured as 16/120. Thus, the throughput is

$$\begin{aligned} 16 \text{ transponders} * 30.3 \text{ Mbps/transponder} &= 484.8 \text{ Mbps} \\ + \ 16 \text{ transponders} * 23.58 \text{ Mbps/transponder} &= 377.28 \text{ Mbps} \\ \text{Total} &= 862.08 \text{ Mbps} \end{aligned}$$

[1] The (146,130) Reed-Solomon code is really a shortened Reed-Solomon code (see Chapter 5).

The 101° W longitude slot can accommodate a fourth satellite, at which time all four of the satellites would be configured as 8/240. At this point, the total throughput would be

$$32 \text{ transponders} * 30.3 \frac{\text{Mbps}}{\text{transponder}} = 969.6 \text{ Mbps}$$

Thus, the Hughes DBS satellites alone will be raining down on CONUS almost one gigabit per second (Gbps).

Since the compression ratio is between 40:1 and 60:1, the backhaul problem is obvious: between 40 Gbps and 60 Gbps must be input to the system every second of every day.

Now we can answer another question about the services that can be provided from a single DBS orbital slot. If an average service requires 4 Mbps, 213 services (channels) can be provided by DIRECTV and USSB now, with the total rising to 242 if a fourth satellite is launched.

3.6 More Services per Orbital Slot

As explained in Chapter 2, the full CONUS orbital slots are extremely valuable. Therefore, there is continuous pressure to increase the number of services that can be provided. Certainly such possibilities as packing the satellites closer together or utilizing satellites with higher frequencies (K_a for example) will enable continuing expansion. Such solutions will not be available in the near future and are currently very expensive, however. The near-term techniques being used to increase the number of services are statistical multiplexing and, ultimately, symbolic compression (MPEG 4).

Statistical Multiplexing (Stat Mux) takes advantage of the fact that the services that are time-division-multiplexed together into a single transponder are statistically independent. Thus, their bursts will happen at different times so that we can allocate bits to each service according to need. Stat Mux will be discussed in more detail in Chapter 10.

Symbolic compression permits the compression of each part of a service according to the service content. Chapter 13 discusses this in detail.

There are certain techniques in Near Video on Demand (NVOD) that can increase the apparent capacity of the system. These involve making the time between service starts smaller while actually reducing the bit rate requirements. These techniques generally require more memory in the IRD. Such techniques will be discussed in Chapter 11.

3.7 Link Analysis

Since a DBS system can make the uplink very powerful, it has a very small effect on the carrier-to-noise ratio. Thus, the downlink carrier-to-noise ratio can be made to depend only on the downlink parameters.

The basic link equation is

$$\left(\frac{C}{N}\right)_d = \text{EIRP}_s \left(\frac{c}{4\pi f_d d_d}\right)^2 \left(\frac{G}{T}\right)_{IRD} \frac{1}{kB} \tag{3.1}$$

where

$\left(\frac{C}{N}\right)_d$ = carrier to noise, on downlink

EIRP_s = satellite EIRP

c = velocity of light = $3 * 10^8$ meters/sec

f_d = downlink frequency

d_d = slant range of downlink, the distance from the satellite to the receiver

$\left(\frac{G}{T}\right)_{IRD}$ = antenna gain to noise temperature ratio for the IRD

k = Boltzman's constant = $1.38 * 10^{-23}$ joules/degrees K

B = noise bandwidth

Equation (3.1) can be simplified by inserting known and fixed values.

First, d_d can be calculated from the geometry as

$$d_d^z = (R_e + H)^2 + R_e^2 - 2R_e(R_e + H) * \sin\left[\Theta + \sin^{-1}\left(\frac{R_e}{R_e + H}\cos\Theta\right)\right] \tag{3.2}$$

where

Θ = elevation angle of IRD receiver

R_e = radius of earth = 6,378 km

H = geostationary orbital height = 37,200 km

$\therefore R_e + H$ = 43,580 km

Let

$$a = \sin\left[\theta + \sin^{-1}\left(\frac{R_e}{R_e + H} * \cos\theta\right)\right] \tag{3.3a}$$

Continuing with the other parts of Equation (3.3a),

$$\sin^{-1}\left(\frac{6.378 * 10^3}{4.358 * 10^4} * \cos\Theta\right) = \sin^{-1}(0.146 * \cos\Theta)$$

Throughout the continental United States, the elevation angle varies from 27° to 54°. For the minimum elevation angle,

$$\sin^{-1}(0.146\cos\Theta) = \sin^{-1}(0.13) = 7.47$$

$$a = \sin\,[27 + 7.47] = 0.566 \tag{3.3b}$$

$$\therefore\ d^2 = (4.358 * 10^4)^2 + (6.378 * 10^3)^2 - 2(6.378 * 10^3)(4.358 * 10^4)$$

where a is defined in (3.3a).

$$d^2 = 18.992 * 10^8 + 40.679 * 10^6 - 55.6 * 10^7 * a$$

Using the value for a from Equation (3.3),

$$d^2 = 18.992 * 10^8 + .41 * 10^8 - 3.15 * 10^8$$

$$d^2 = 16.26 * 10^8\text{, or} \tag{3.4}$$

$$d = 40{,}324\ \text{km}$$

At the maximum elevation angle,

$$\sin^{-1}(0.146\cos\Theta) = \sin^{-1}(0.0858) = 4.92$$

$$a = \sin\,[54 + 4.92] = .856$$

$$d^2 = 18.992 * 10^8 + 40.679 * 10^6 - 55.6 * 10^7 * a \tag{3.5}$$

$$d^2 = 18.992 * 10^8 + .41 * 10^8 - 4.76 * 10^8 = 14.65 * 10^8$$

$$\therefore\ d = 38{,}257\,\text{km}$$

The entity $\left(\frac{c}{4\pi f_d}\right)^2 * \left(\frac{1}{d_d}\right)^2$ from Equation (3.2) then can be calculated (using $f_d = 12.45 \times 10^9$ as the average downlink frequency)

$$\left(\frac{c}{4\pi f_d}\right)^2 * \left(\frac{1}{d_d}\right)^2 = \left(\frac{3 * 10^8}{4\pi * 12.45 * 10^9}\right)^2 * \left(\frac{1}{d_d}\right)^2 = 3.68 * 10^{-6} * \left(\frac{1}{d_d}\right)^2$$

Finally, the constant

$$k * B = 1.38 * 10^{-23} * 24 * 10^{6} \frac{\text{joules}}{\text{degrees } K * \text{sec}}$$

$$= 3.312 * 10^{-16} \frac{\text{joules}}{\text{degrees } K * \text{sec}}$$

$$\frac{1}{k * B} = 3.02 * 10^{15} \frac{\text{degrees } K * \text{sec}}{\text{joule}} \tag{3.6}$$

$$\therefore \left(\frac{c}{4\pi f_d}\right)^2 \frac{1}{k * B} * \left(\frac{1}{d_d}\right)^2 = 3.68 * 10^{-6} * 3.02 * 10^{15} * \left(\frac{1}{d_d}\right)^2$$

$$= 1.11 * 10^{10} * \left(\frac{1}{d_d}\right)^2$$

For the minimum elevation angle, $d_d = 4.0324 * 10^7$ meters, so $(1/d_d)^2 = 6.15 * 10^{-16}$, and Equation (3.6) becomes $6.83 * 10^{-6}$. For the maximum elevation, $d_d = 3.8275 * 10^7$ meters, so $(1/d_d)^2 = 6.826 * 10^{-16}$ and (3.6) becomes $7.58 * 10^{-6}$.

For the minimum elevation angle,

$$\left(\frac{C}{N}\right)_d = EIRP_S * \left(\frac{G}{T}\right)_{IRD} * 6.83 * 10^{-6}, \tag{3.7}$$

and for the maximum elevation angle,

$$\left(\frac{C}{N}\right)_d = EIRP_S * \left(\frac{G}{T}\right)_{IRD} * 7.58 * 10^{-6}. \tag{3.8}$$

Equation (3.7) will always give the lower result, and (3.7) and (3.8) will be very close in any event, so we will continue the calculation only for Equation (3.7).

For the DBS satellites, EIRP is 53 dBW to 56 dBW, so

$$\therefore \left(\frac{C}{N}\right)_{dlow} = 2 \times 10^5 * \left(\frac{G}{T}\right)_{IRD} * 6.83 \times 10^{-6} = 1.4678\left(\frac{G}{T}\right)_{IRD}$$

$$\left(\frac{C}{N}\right)_{dlow,\,\text{dB}} = 1.667 + 10\log_{10}\left(\frac{G}{T}\right)_{IRD}$$

$$\therefore \left(\frac{C}{N}\right)_{dhigh} = 2.9356\left(\frac{G}{T}\right)_{IRD}$$

or

$$\left(\frac{C}{N}\right)_{dhigh,\,\text{dB}} = 4.667 + 10\log_{10}\left(\frac{G}{T}\right)_{IRD}$$

$$G = \eta\left(\frac{\pi f D}{c}\right)^2,$$

where η is a combination of losses, $\eta = \eta_1\eta_2\eta_3\eta_4\eta_5\eta_6$. When typical values are used, $\eta \cong .5$.

Since the U.S. DBS systems use an 18-inch (0.45-meter) diameter dish,

$$\therefore G = 0.5 \frac{\pi * 12.45 * 10^9 * 0.45^2}{3 * 10^8}$$

$$= 0.5\left(\frac{17.601 * 10^9}{3 * 10^8}\right)^2$$

$$= 0.5\left(5.867 * 10^1\right)^2 = 17.2 * 10^2$$

$$G_{\text{dB}} = 10\log_{10}(17.2 * 10^2) = 32.355 \text{ dB}$$

A typical $T = 125°\text{K}$

$$\Rightarrow \frac{G}{T} = 11.39 \text{ dB}/°\text{K}$$

$$\left(\frac{C}{N}\right)_{dlow,\,\text{dB}} = 13.05 \text{ dB}$$

$$\left(\frac{C}{N}\right)_{dhigh,\,\text{dB}} = 16.05 \text{ dB}$$

It will be shown in Chapter 5 that the threshold C/N is about 5 dB[2] for the low-power case. Thus, there is a margin of more than 8 dB. This might

[2] The requirement is really on E_b/N_O, but for the low-power case, C/N and E_b/N_O are almost identical.

seem to be a rather substantial overdesign, except for the fact that the DBS systems, located in K_u-Band, must deal with rain attenuation.

3.8 Degradation Because of Rain

Above 10 GHz, electromagnetic waves are attenuated by rain. Rain impairs performance in three ways:

1. Attenuation of the desired signal
2. Increase in sky noise temperature
3. Interference from depolarization of an opposite polarization transponder if frequencies are reused as in DBS

Each of these effects must be evaluated separately and the overall effect on carrier-to-noise ratio evaluated. Because of the complexity, these calculations are contained in Appendix A.

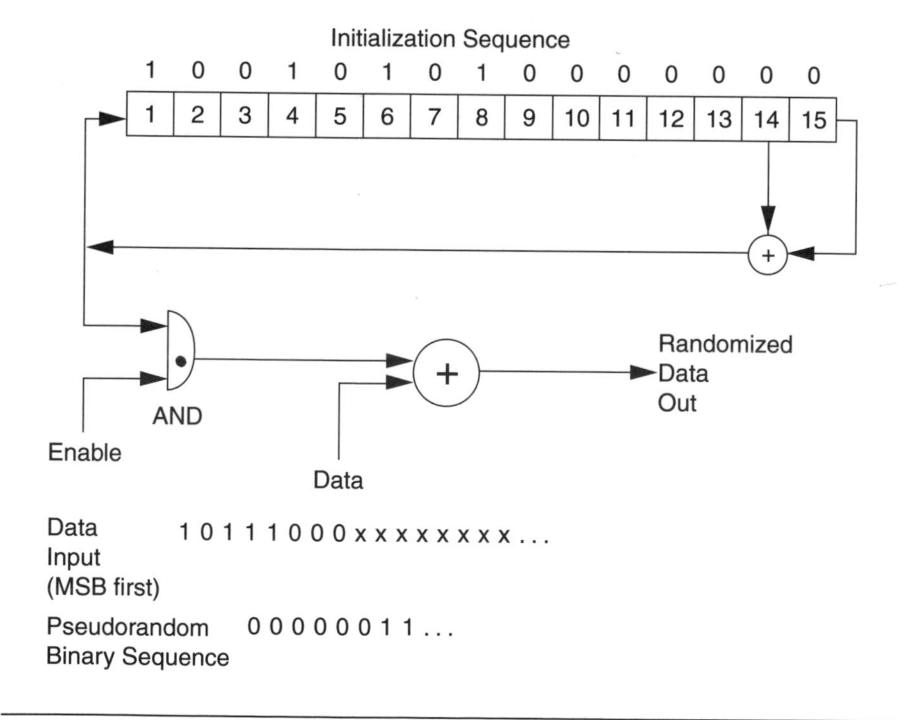

Figure 3.4 Energy Dispersal Circuit for DVB

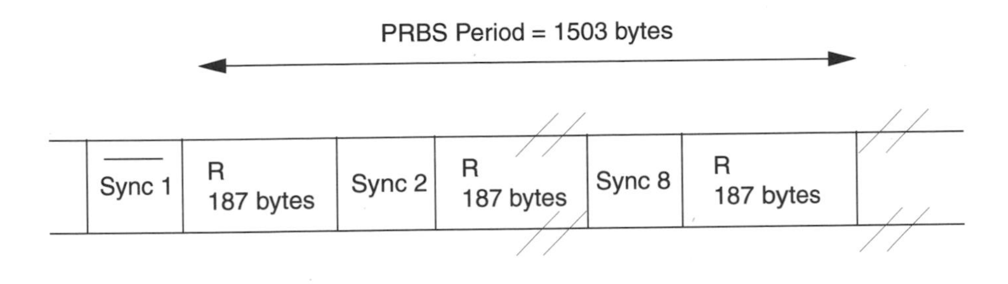

Figure 3.5 Energy Dispersal Framing for DVB

3.9 Energy Dispersal

If the transport packets have long runs of 1s or 0s, it becomes difficult or impossible to synchronize the receiver with the incoming bitstream. Thus, there is a step in the processing chain that disperses energy, to ensure that there are no long strings of the same bit type.

This operation is performed on the transport stream before any subsequent processing (FEC, modulation, and so on). In the receiver, the output of the outer (Reed-Solomon) decoder is the input to an energy-dispersal inverse, which then delivers transport packets to the downstream parts of the receiver.

Figure 3.4 shows the circuit that is used in the DVB Standard for energy dispersal. The generator polynomial for the Pseudo Random Binary Sequence (PRBS) is

$$g(x) = 1 + X^{14} + X^{15}$$

as can be seen in Figure 3.4.

Note that the first byte is the logical complement of the MPEG 2 Systems sync byte. In other words, the MPEG 2 Systems sync byte is 47 hex. The one's complement of this is B8 hex. This is transmitted before the PRBS is started. The PRBS then runs for the next 187 bytes of the first packet and the next seven packets for a total of 1,503 bytes. Figure 3.5 shows this framing structure.

Note that the circuit in Figure 3.4 is symmetric. In the transmitter, the data input is from the TDM multiplexer. In the receiver, the data is from the Reed-Solomon decoder.

References

[ETSI95] *ETSI,* "Digital Video Broadcast-Satellite," European Technical Standards Institute Standard ETSI 300–421, 1995.

[Ha90] Ha, Tri, *Digital Satellite Communication,* New York: McGraw-Hill, 1990.

[Ippolito86] Ippolito, Louis, *Radiowave Propagation in Satellite Communication,* New York: Van Nostrand Reinhold, 1986.

[Roddy95] Roddy, Dennis, *Satellite Communications,* 2nd ed., New York: McGraw-Hill, 1995.

PART TWO

Key Subsystems

CHAPTER FOUR

Key Elements of the Radio Frequency Subsystem

4.1 Introduction

Many of the components shown in Figure 3.1 are standard in the satellite communication business and will not be discussed in this book. However, the shaped reflector antenna, modulation (important for establishing the overall DBS performance), Low Noise Block (LNB), and Traveling Wave Tube Amplifier (TWT-A) are key radio frequency (RF) subsystems and will be discussed individually in this chapter.

4.2 The Shaped Reflector Antenna

Appendix A shows that the rain-fades vary widely throughout the United States. This means that if nothing special were done, areas receiving high rainfall would have unacceptable service because of numerous service outages.

One of the early ideas was to use larger antennas in the higher rainfall areas. However, with a population mobility of 10 percent in the United States, someone from a dry area might move to a wet area and would be very unhappy because his or her service quality would go down. Furthermore, having multiple size antennas would create manufacturing complications. The solution was the highly innovative shaped reflector antenna.

Most of us have gone to an amusement park Fun House where oddly shaped mirrors distort our appearance. The concept behind the shaped antenna reflector is similar: the reflector is shaped to create more energy in some parts of the beam and less energy in other parts.

Before the invention of the shaped reflector, antenna designers had to shape the transmitted beam by an arrangement of feed horns. These feed horns had to be excited in proper phase and amplitude by a complex beam-forming network.

As customer requirements have grown, so too has the complexity of the beam-forming network and the horn arrays. The increased complexity adds mass to the satellite. This increases the launch cost from $25,000 to $50,000 per pound. The early DIRECTV antenna design had about 120 feed horns, each of which had to be fine-tuned to maintain proper amplitude and phase.

Mechanically shaping the reflector eliminated the costly feeds and the undesired extra mass. Now, the dimpling in the reflector surface substitutes for the complex multiple feed arrays in directing radiated energy in the right directions. Antenna designers have eliminated the need for all but one horn and have done away with the feed network entirely.

The remaining horn illuminates the shaped reflector to produce the desired beam coverage. The absence of the horn arrays and the beam-forming network reduces signal losses and improves antenna performance.

Analysis shows that shaped-reflector technology significantly increases its effective radiated power. Compared to a more traditional multiple-feed antenna system, the new technology reduces the weight of the antenna systems for the DBS satellites by approximately 250 pounds. Figure 4.1 shows a microphotograph of the reflector surface and a sketch of how the shaping affects the beam formation.

> *Note:* There is an additional benefit to the shaped reflector antenna that doesn't affect the DBS communication system. Feed arrays usually have to be physically offset from the spacecraft structure to prevent blockage or scattering of the radiated signal from interfering with the antenna pattern. Offsetting the feed horns and feed assembly shifts the satellite's center of gravity. This has to be corrected by a time-consuming process of balancing the satellite for a successful launch. Feed weight on satellites equipped with a shaped reflector antenna is significantly decreased. Therefore, the process of balancing the spacecraft is greatly simplified [Vectors93].

This shaping of the reflector is computed in advance according to the rainfall model for the United States to normalize the expected outage time for all areas. The probability is 0.997 that the outage will be less than 40 hours per year.

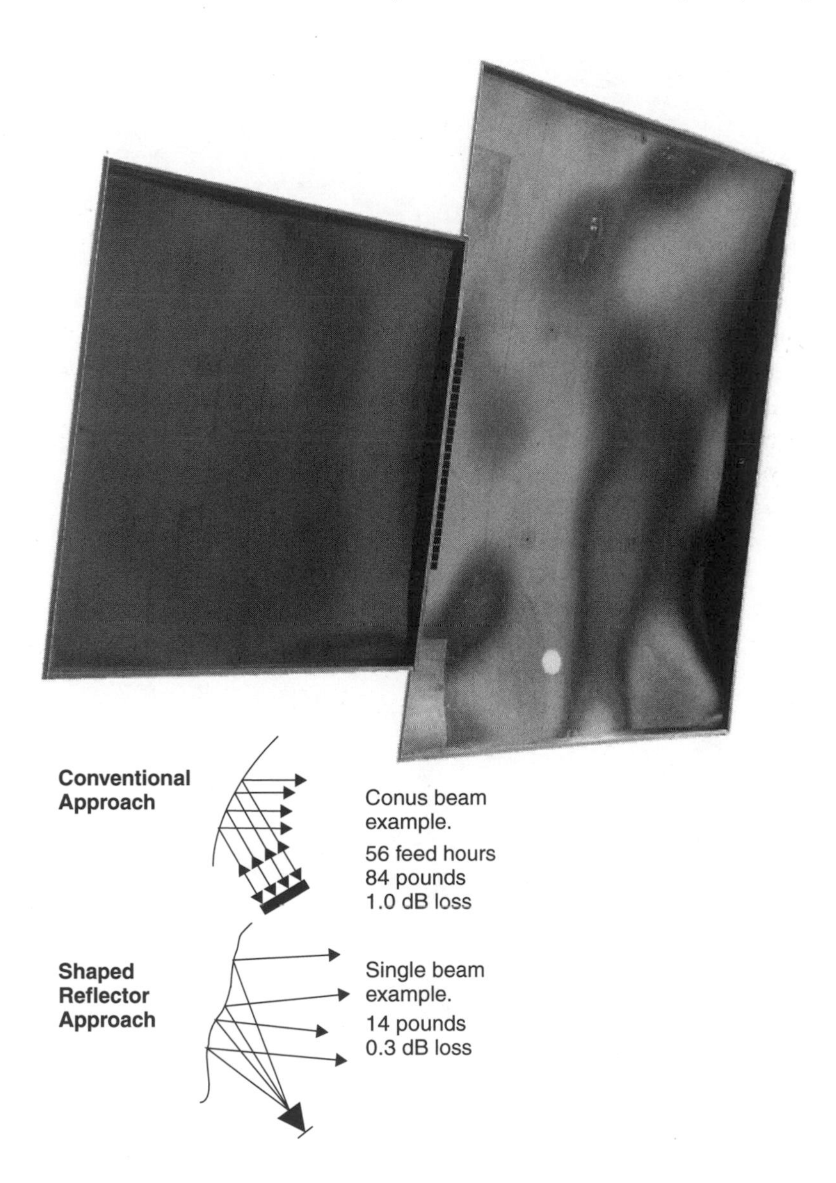

Figure 4.1 The Shaped Reflector Antenna Reproduced from *Vectors*, XXXV(3):14, 1993. Copyright © 1993 Hughes Aircraft Co. All rights reserved. Used with permission.

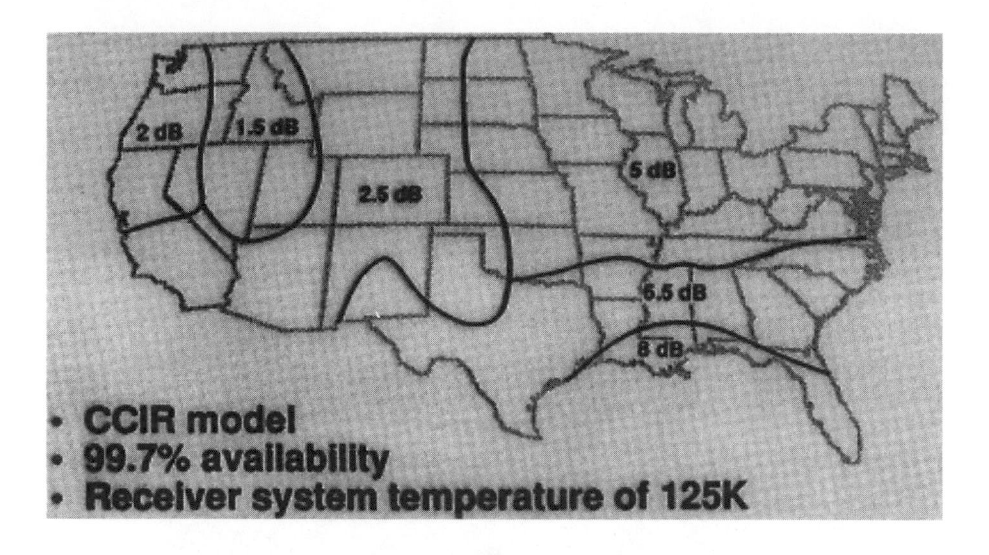

Figure 4.2 Required Rain Margins for the United States Reproduced from "Propagation in Nonionized Media," ITU-R Rec. PN.837-1, 1994 PN Series. Used with permission of ITU, Geneva, Switzerland.

Figure 4.2 shows a relative energy given to various geographic areas in the United States. This figure can be derived from the rain model given in Appendix A and the DBS parameters.

4.3 Modulation and Demodulation

Modulation refers to the way that information is imparted to an RF carrier. In analog communication systems, the modulation is amplitude, frequency, or phase, and these entities are allowed to vary continuously. In a digital communication system, the information is 1s and 0s; however, we cannot radiate 1s and 0s directly. Therefore, the 1s and 0s are used to modulate a carrier or sequence of carriers. In this digital modulation, only a finite number of states are possible each symbol time. The problem at the receiver is to determine which of the states was transmitted.

The simplest way to do this digital modulation is to represent a 1 by the cosine $(\omega_c t)$ and a 0 by –cosine $(\omega_c t)$. This is shown in Figure 4.3. Also shown in this figure is a decision boundary that corresponds to the y-axis. If the received signal falls anywhere to the left of this boundary, the signal is called

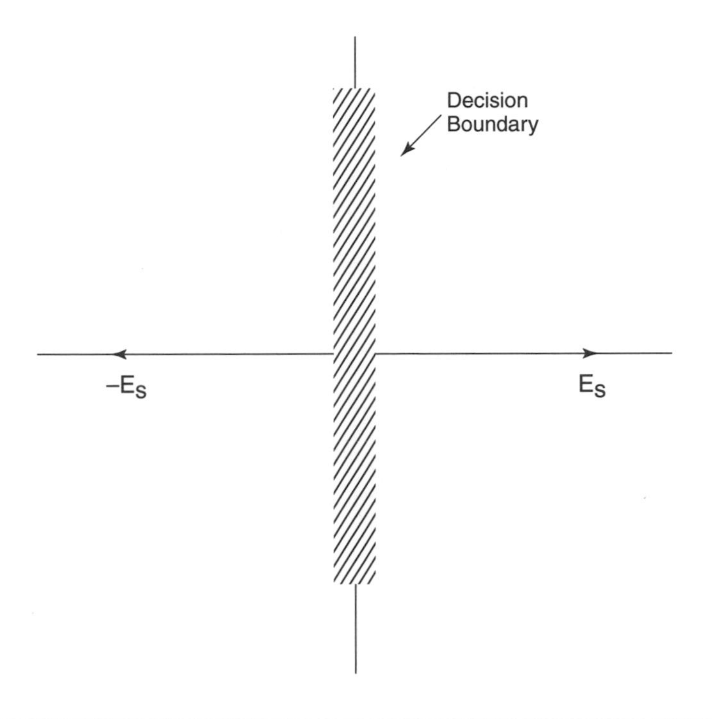

Figure 4.3 Binary Phase Shift Keying

a 0, otherwise it is called a 1. This type of modulation is called Binary Phase Shift Keying (BPSK). The time it takes each of the two possibilities to be transmitted is called a symbol time. One bit of information is transmitted during this symbol time.

However, we can convey more than one bit of information per symbol. Perhaps the most popular way to do this is with QPSK. With QPSK, the carrier can assume one of four possible phases, as shown in Figure 4.4. Each of the carrier positions can be described by cosine $[w_t * t + j * \pi/2], j = 0, 1,2,3$. It should be clear that each position of the carrier represents two bits. Thus, QPSK permits the transmission of two bits per symbol, or twice as much information per symbol as BPSK. Appendix B shows how QPSK modulators and demodulators can be implemented.

It should be clear that we can continue this process by using 8 carrier states (8-ary), which yields three bits per symbol, or even using 16 carrier states (16-ary), which gives four bits per symbol. However, the hardware

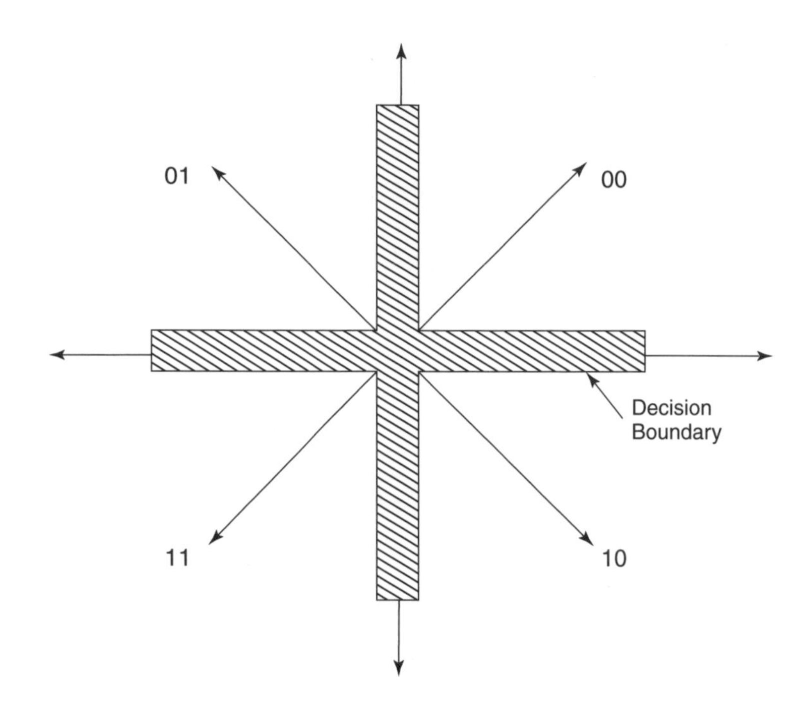

Figure 4.4 Quaternary (Quadrature) Phase Shift Keying

complexity increases and, worst of all, the required antenna diameters increase significantly.

4.4 The Low-Noise Block

As noted in Chapter 2, the downlink frequency for DBS is 12.2 to 12.7 GHz. At this frequency, signal degradation is sufficiently high so that attempts to put the signal on any length of coaxial cable will not work. Thus, a way has to be found to lower the frequency before it is sent from the Outdoor Unit (ODU) to the Indoor Unit (IDU). This function is performed by the LNB.

In Figure 4.5, the LNB is the device that is located at the offset focus of the reflector. Figure 4.6 is a block diagram of the LNB. Starting from the left of the figure, the low-noise amplifier amplifies the received signal. This signal is then multiplied by a signal from the local oscillator (lo), which operates at a frequency of 12,262 MHz. Since the $\sin(\omega_s t) * \sin(\omega_{lo} t) = \sin\{(\omega_s + \omega_{lo})t\} + \sin\{(\omega_s - \omega_{lo})t\}$ the output of the mixer contains the sum and the difference of the two frequencies.

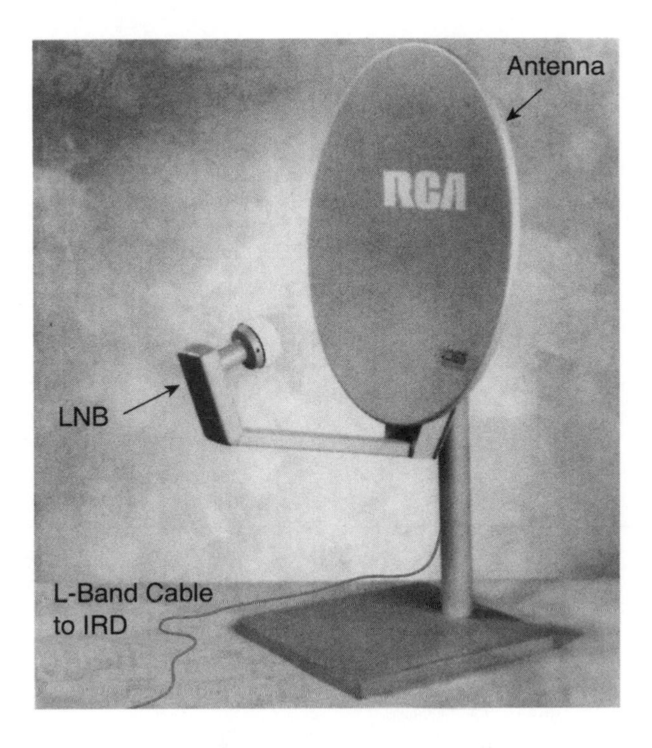

Figure 4.5 LNB Mounted on Antenna (Outdoor Unit)

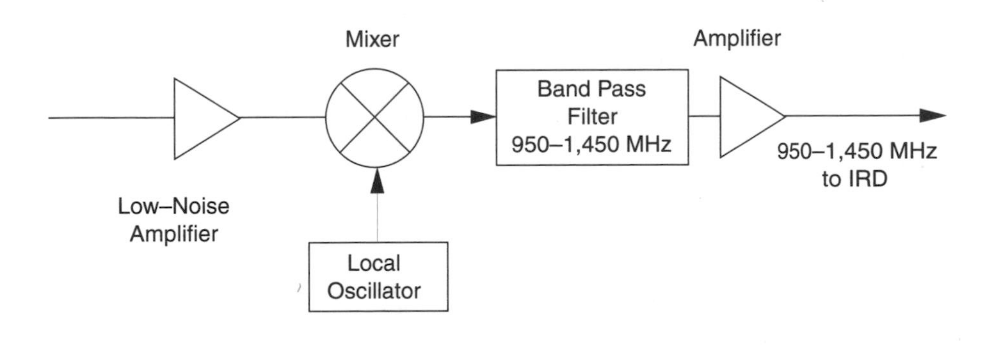

Figure 4.6 LNB Block Diagram

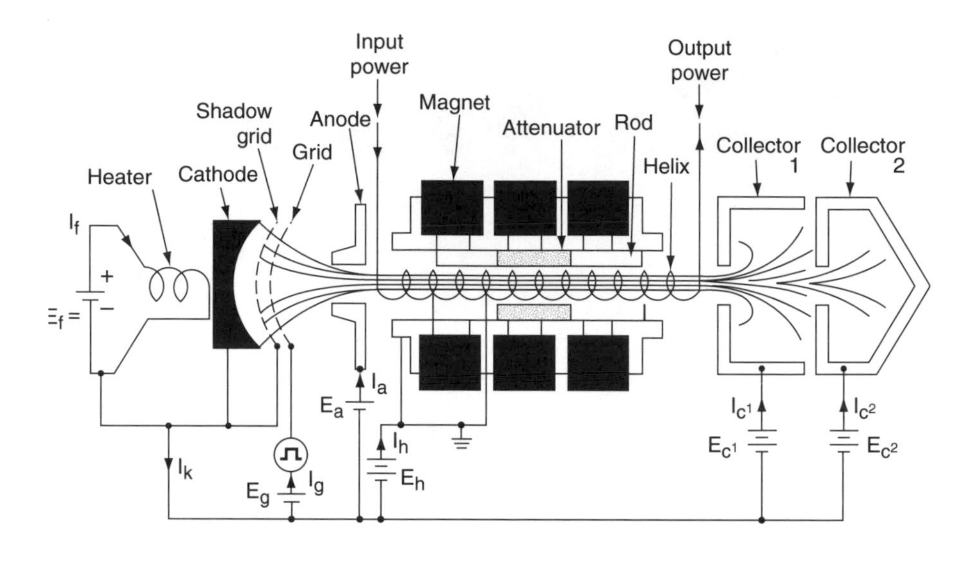

Figure 4.7 Traveling Wave Tube Schematic Reproduced from *TWT/TWTA Handbook,* p. 66. Copyright © 1992, Hughes Aircraft Co. Used with permission.

The local oscillator frequency was selected such that the difference frequency would fall in the range of 950 MHz to 1,450 MHz (L-Band). A bandpass filter selects this difference frequency, which is then amplified by the distribution amplifier, that drives the coaxial cable from the LNB to the set-top box (IRD).

4.5 Traveling Wave Tube Amplifier

Traveling wave tube amplifiers are used for the DBS satellite high-power amplifiers. The DBS satellites use 120-watt TWT-As. Two of them can operate in parallel to create the 240-watt transponder mode.

Figure 4.7 is a schematic of a TWT-A. The heater, cathode, focusing electrodes, and collectors form an electron beam. The RF signal to be amplified is coupled into a helix through which the electron beam passes. A traveling wave is set up along the helix. This traveling wave interacts with the electron beam, which converts energy from the electron beam into the RF signal. Thus, the RF output signal is an amplified version of the input. Power from the solar panel (or batteries) provides DC power for the electron beam and hence to the RF power out. The efficiency of this process is 65 percent.

References

[Ha90] Ha, Tri, *Digital Satellite Communication,* New York: McGraw-Hill, 1990.

[Hughes92] *TWT and TWTA Handbook,* Hughes Aircraft Company, Torrance, CA, 1992.

[Hughes93] Hughes Electronics, *Vectors,* XXXV(3), Hughes Aircraft Company, Torrance, CA, 1993.

CHAPTER FIVE

Forward Error Correction

All digital communication systems suffer from errors. An error is made if a transmitted 0 is decoded as a 1, or vice versa. DBS systems are particularly vulnerable to transmission errors because they are distributing highly compressed files. Thus, the effect of a bit error can be very large.

5.1 What Is Error Correction?

To combat bit errors, a number of techniques have been developed, including forward error correction (FEC). Consider Table 5.1a. The column to the right

Table 5.1a The Concept of an Error-Correcting Code

Message	2^3	2^2	2^1	2^0	Row Parity
1	0	0	0	0	0
2	0	0	0	1	1
3	0	0	1	0	1
4	0	0	1	1	0
5	0	1	0	0	1
6	0	1	0	1	0
7	0	1	1	0	0
8	0	1	1	1	1
9	1	0	0	0	1
10	1	0	0	1	0
11	1	0	1	0	0
12	1	0	1	1	1
13	1	1	0	0	0
14	1	1	0	1	1
15	1	1	1	0	1
16	1	1	1	1	0
Column Parity	0	0	0	0	

of the vertical line is a parity on the bits to the left of it. The value is a 1 if there is an odd number of 1s, and a 0 if there is an even number of 1s. Likewise, the bits below the horizontal line are parity bits on the columns. The entries in the table are the original information bits. Now, instead of sending only the information bits, we send both the information bits, the right column, and the last row. Then we can correct any single bit error that is made in transmission, including the parity bits themselves. In certain cases more than one bit error also can be corrected.

To see this, suppose that the 2^1 bit of message 7 is decoded incorrectly as a 0, as shown in Table 5.1b. When the decoder forms the equivalent of Table 5.1b, it will recognize that row 7 and the 2^1 column are wrong because they have an odd number of 1s and the parity bit indicates that they should have an even number of 1s. The intersection of message 7 and the 2^1 column is the 2^1 bit of message 7, which is then known to be in error and can be corrected.

You should experiment with Table 5.1a, making any single-bit error in the message bits or even a parity bit. A few tries should convince you that all single-bit errors can be corrected.

Table 5.1b Results When Error in Third Bit of Message 7 of Table 5.1a Is Received

Message	2^3	2^2	2^1	2^0	Row Parity
1	0	0	0	0	0
2	0	0	0	1	1
3	0	0	1	0	1
4	0	0	1	1	0
5	0	1	0	0	1
6	0	1	0	1	0
7	0	1	**0**	0	0
8	0	1	1	1	1
9	1	0	0	0	1
10	1	0	0	1	0
11	1	0	1	0	0
12	1	0	1	1	1
13	1	1	0	0	0
14	1	1	0	1	1
15	1	1	1	0	1
16	1	1	1	1	0
Column Parity	0	0	0	0	

5.1.1 Code Rate

It should be clear from the preceding example that if this type of code is used to reduce the bit error rate, more bits will have to be transmitted than if only the information bits are transmitted. This is expressed in a digital communication system by the code rate, which is defined as the information bits per second divided by the total bits per second. Typical code rates vary from one-half to seven-eighths, although recent work may be able to take advantage of code rates lower than one-half.

In the code depicted in Tables 5.1a and 5.1b (sometimes called Elias codes [Biglieri+91]), the code rate is 64/84 = .762. Note that there are two ways to deal with the parity bits: the bits per second can be increased, or the total bits per second can be fixed and the information bit rate decreased. DBS systems are always bandwidth-limited, so the latter approach is used.

5.2 Types of Codes

There are basically two types of codes: block codes and convolutional codes. First, these two types of codes will be examined separately. Next, combinations of block and convolutional codes, called concatenated codes, will be examined.

5.2.1 Block Codes

A block code is a mapping of k input symbols into n output symbols, with $n > k$. The code is called an (n, k) code. The n symbol block is called a code vector. The k symbol block is called the message, or message vector.

It will be shown that, without any loss of code capability, the first k symbols of a code vector can be the message symbols. This type of code is said to be systematic. Thus, $n - k$ symbols are parity symbols. The code rate of an (n, k) code is k/n.

Therefore, a code vector can be written as

$$c = i_1 i_2 \ldots i_k p_1 p_2 \ldots p_r$$

where i_k are information symbols and the p_1 are parity symbols, $n = k + r$.

In later sections we need the algebraic definition of a *finite* field. A field has two operations between its elements, called $(\cdot)$ and $\oplus$, which must satisfy certain requirements. When a field contains only the elements 0 and 1, modulo-two arithmetic defined by

(•)

a	b	out
0	0	0
0	1	0
1	0	0
1	1	1

and

⊕

a	b	out
0	0	0
0	1	1
1	0	1
1	1	0

forms a finite field of two elements. Note that (•) is an AND circuit and ⊕ is logically an exclusive OR circuit. For the remainder of this section we concentrate on binary fields.

To provide error detection and correction, the parity bits must be formed from the information bits. This can be done by forming

$$\begin{aligned} P_1 &= Z_{11}\mathbf{i}_1 \oplus Z_{12}\mathbf{i}_2 \oplus \cdots \oplus Z_{1k}\mathbf{i}_\mathbf{k} \\ P_2 &= Z_{21}\mathbf{i}_1 \oplus Z_{22}\mathbf{i}_2 \oplus \cdots \oplus Z_{2k}\mathbf{i}_\mathbf{k} \\ &\vdots \\ P_\tau &= Z_{\tau 1}\mathbf{i}_1 \oplus Z_{\tau 2}\mathbf{i}_2 \oplus \cdots \oplus Z_{\tau k}\mathbf{i}_\mathbf{k} \end{aligned} \tag{5.1}$$

where ⊕ denotes exclusive OR and the product of Z_{pq} and i_q, $Z_{pq}i_q$ is (•). In matrix notation, this becomes

$$[\mathbf{P}] = [\mathbf{Z}]\,[\mathbf{i}] \tag{5.2}$$

Physical Interpretation of Z

If the information (message) bits are in a shift register and the parity bits are formed by exclusive OR circuits whose inputs are from the stages of the shift register, then **Z** is the connection diagram that indicates which stages of the shift register are connected to which exclusive ORs. A 1 indicates a connection, and a 0 no connection.

If we form the matrix

$$\mathbf{G} = [\mathbf{I}_{k*k} \vdots \mathbf{Z}^t] \tag{5.3}$$

where t = matrix transpose, we can write

$$C = \mathbf{m}[\mathbf{G}] \tag{5.4}$$

where **m** is the message vector and **G** is called the generator matrix.

Example 5.1 The generator matrix for the (7, 4) Hamming code, which will be discussed later in the Hamming Codes section, is

$$\mathbf{G} = \begin{bmatrix} 1 & 0 & 0 & 0 & 1 & 1 & 0 \\ 0 & 1 & 0 & 0 & 1 & 0 & 1 \\ 0 & 0 & 1 & 0 & 0 & 1 & 1 \\ 0 & 0 & 0 & 1 & 1 & 1 & 1 \end{bmatrix}$$

If the message is $\mathbf{m}$ = 1011, then

$$C = [\mathbf{m}][\mathbf{G}] = [1011010]$$

Note that the first four bits of $C = \mathbf{m}$, so the code is systematic.

Decoding

The decoder receives an n-bit code vector $\tilde{\mathbf{C}}$. This is written $\tilde{\mathbf{C}}$ because $\tilde{\mathbf{C}}$ may contain bit errors. The problem of the decoder then is to recover **m** from $\tilde{\mathbf{C}}$. This is accomplished by calculating the row vector **S** called the syndrome[1] as

$$\mathbf{S} = \tilde{\mathbf{C}}\mathbf{H}^{\mathrm{T}} \tag{5.5}$$

where H is called the parity check matrix. If $\mathbf{S} = 0$, no errors are detected. If $\mathbf{S} \neq 0$, there is one or more errors. **H** is formed as

$$\mathbf{H} = [\mathbf{Z}\mathbf{I}_{r*r}] \tag{5.6}$$

which is $r * n$.

[1] Couch [97] notes that in medicine, the word *syndrome* is a pattern of symptoms. In this case, the disease consists of bit errors.

Example 5.2 Assume the same conditions as in Example 5.1. **H** then can be formed as

$$\mathbf{H} = \begin{bmatrix} 1 & 1 & 0 & 1 & 1 & 0 & 0 \\ 1 & 0 & 1 & 1 & 0 & 1 & 0 \\ 0 & 1 & 1 & 1 & 0 & 0 & 1 \end{bmatrix} \tag{5.7}$$

Suppose the message $\mathbf{m} = [1011]$ as in Example 5.1 so $\tilde{\mathbf{C}} = [1011010]$. Now suppose that $\tilde{\mathbf{C}} = [1010010]$; that is, the middle bit has been corrupted from a 1 to a 0. Then, we form $\mathbf{S} = \tilde{\mathbf{C}}\mathbf{H}^T$ or

$$\mathbf{S} = [1010010] \begin{bmatrix} 1 & 1 & 0 \\ 1 & 0 & 1 \\ 0 & 1 & 1 \\ 1 & 1 & 1 \\ 1 & 0 & 0 \\ 0 & 1 & 0 \\ 0 & 0 & 1 \end{bmatrix} = [111] \tag{5.8}$$

Because $\mathbf{S} \neq 0$, bit errors have occurred.

To determine the location, we compare **S** to the columns of **H**. Because **S** equals the middle column of **H**, this is the location of the bit error. Therefore, we know $\mathbf{C} = 1011010$ and $\mathbf{m} = 1011$.

Geometric Interpretation

Block codes can be given a geometric interpretation. Imagine the generalization of a sphere to n-dimensional space from the normal three-dimensional space. A sphere in three-dimensional space is called a hyper sphere in n space. Consider a number of such spheres packed in a box. Let the center of each sphere be a code vector. A number of possible received vectors fall in the sphere but are not the center. All such vectors are mapped to the code vector, which is at that sphere's center.

Construction of a code is like packing the spheres in a box. If we make the spheres small, we can get more of them in the box (higher code rate). The probability of error is larger, however, because a received vector is more likely to fall in an incorrect sphere when it is corrupted by noise. Making the spheres larger lowers the code rate but increases the error protection.

Note that there is volume in the box not filled by any sphere. There are a number of possible decision strategies for these received vectors. Perhaps the most obvious is to declare them detected but not corrected errors.

Hamming Codes

Hamming codes are a class of codes that can correct all single-bit errors. The Hamming weight of a binary code vector is defined as the number of 1s. The Hamming distance d between two code vectors is the number positions where they are different. If s is the number of bits that can be detected and t is the number of bits that can be corrected,

$$d \leq s + t + 1 \tag{5.9}$$

Hamming codes are binary block codes with a Hamming distance of 3. Since $d \leq 2t + 1, t = 1$ and 1 error can be corrected.

Hamming codes are easy to construct. The H matrix is determined by choosing the Z matrix within H such that the columns of H contain all of the possible r-bit binary words except for the all-zero word. Only certain combinations of n and k can be Hamming codes. These are of the form

$$(2^m - 1, 2^m - 1 - m), m \geq 3$$

The first five Hamming codes are thus

m	n	k
3	7	4
4	5	11
5	31	26
6	63	57
7	127	120

Because R approaches 1 as n becomes large, these codes become very efficient, although only a single bit error can be corrected.

Other Block Codes

There are a number of types of important block codes. Perhaps the most important are the cyclic codes, which include the Bose Chaudhuri Hoequemhem (BCH), and Reed-Solomon codes, which will be discussed later in this chapter.

5.2.2 Convolutional Codes

The encoders for block codes have no memory. That is, a k-bit message vector is input to the encoder, and the encoder outputs an n-bit code vector.

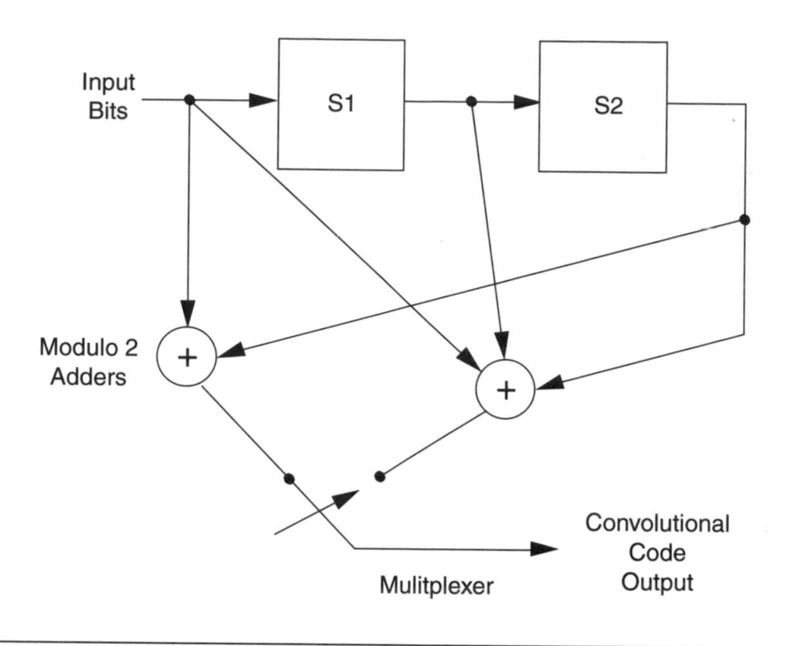

Figure 5.1 Rate 1/2 Convolutional Encoder with Constraint Length = 3

When a sequence of message vectors is input to the encoder, there is no relationship between the output successive code vectors.

Convolutional encoders, on the other hand, have memory. Each time k bits are shifted in (one input frame), n bits are shifted out (one output frame), $n > k$. An input frame influences the output frames for *constraint length* (CL) input frames (CL > 1).[2]

The basic concept of convolutional code is that an input bit will influence a number of output bits. Thus, if the channel corrupts some of the output bits, the input bits still can be reconstructed in the decoder. Figure 5.1 shows a simple convolutional encoder. Each time a bit is moved into the shift register, the multiplexer outputs two bits, one from each of the two modulo-2 adders.

The operation of the encoder can be described by the code tree shown in Figure 5.2 [Couch97]. At each branch, the code takes the upper branch if the input is a 0 and the lower if the branch is a 1. The output for each input is shown in parentheses above the branch. These outputs assume the shift register starts in the all-zeros condition.

[2] The literature on convolutional coding has a number of somewhat different definitions of constraint length. In this book we only consider single-shift register convolutional codes. In this case, the constraint length is the number of stages of the shift register plus one.

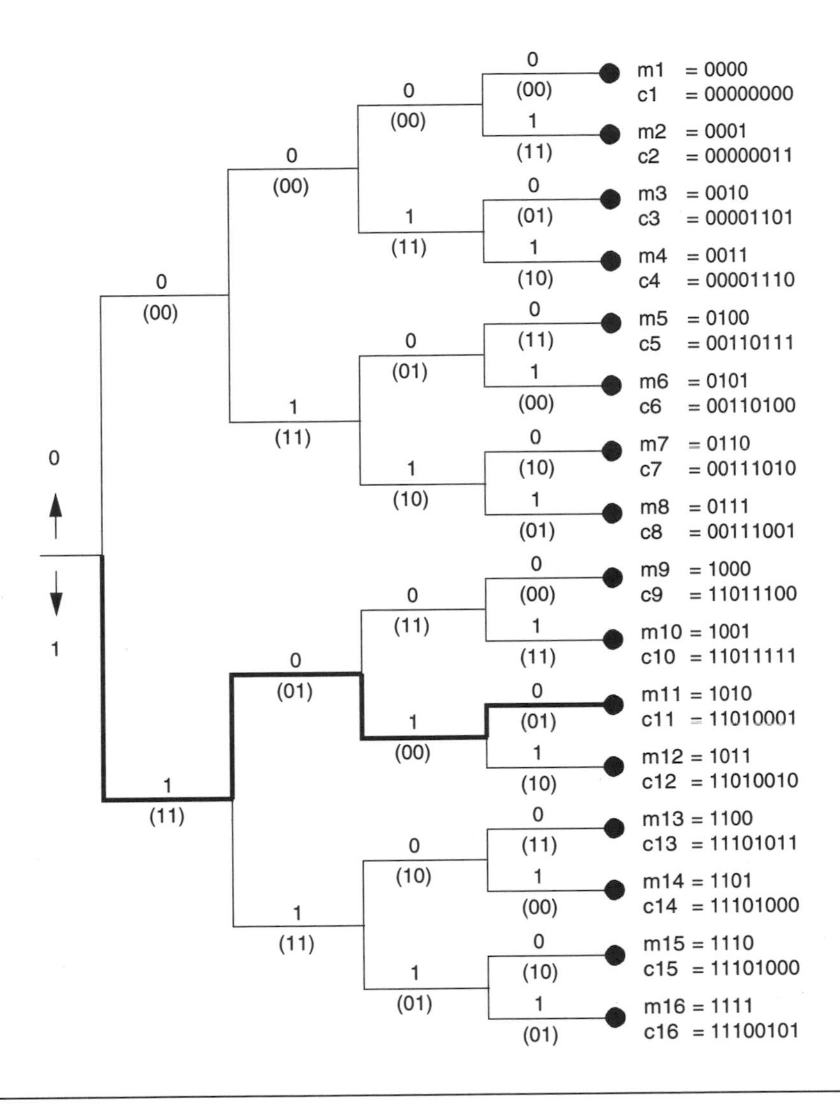

Figure 5.2 Code Tree for (2,1,3) Convolutional Code From (Couch97), p. 30. Copyright © Prentice-Hall, Inc. Reprinted by permission of Prentice-Hall, Inc., Upper Saddle River, NJ.

The convolutional code is decoded by matching the encoded data to the corresponding pattern in the code tree. For example, if $c = 11010010$ is received, Path A of Figure 5.2 is an exact match and $m = 1010$ is the output. If noise corrupts bits of the received vector, there will not be an exact match. In this case, one strategy for selecting the optimum path of the coding tree is to select the path that has the minimum Hamming distance to the received vector. However, as code length is increased, the complexity increases exponentially.

An optimum decoding algorithm, called Viterbi decoding, examines the possible paths and selects the best one based on conditional probabilities. Optimum Viterbi decoding uses soft decisions, where the decoder outputs a result and confidence number indicating how close the decision was to the threshold. It can be shown that soft decision decoding can lead to a 2-dB reduction is the required E_b/N_o. Section 5.6 discusses convolutional codes and Viterbi decoding in more detail.

5.2.3 Code Interleaving

Individual block and convolutional codes perform best when the bit errors are randomly distributed and not correlated with each other. However, in a number of practical cases the errors occur in bursts. Some examples include ignition noise, telephone channels with switching noise, and so on. One important example for DBS occurs when the convolutional coder operates near threshold. In this case, its uncorrected errors occur in bursts.

Interleaving can ameliorate the effects of these burst errors. To understand interleaving, consider the coded bitstream written into a random access memory (RAM) by columns and then read out by rows, as shown in Figure 5.3. In the receiver, the received bitstream first is deinterleaved by the inverse of the process shown in Figure 5.3 and then decoded.

(a) Writing Order

1	6	11	16	21	26	31	36
2	7	12	17	22	27	32	37
3	8	13	18	23	28	33	38
4	9	14	19	24	29	34	39
5	10	15	20	25	30	35	40

(b) Read Out Order

1	2	3	4	5	6	7	8
9	10	11	12	13	14	15	16
17	18	19	20	21	22	23	24
25	26	27	28	29	30	31	32
33	34	35	36	37	38	39	40

Figure 5.3 Simple Interleaving

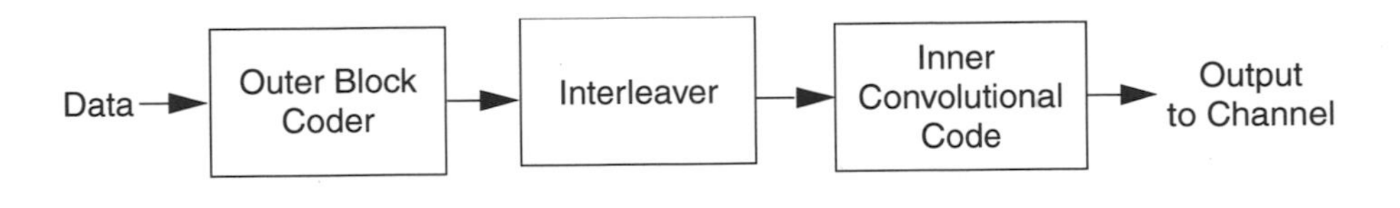

Figure 5.4 Concatenated Code Encoder

Suppose a noise burst corrupts received bits 17, 18, and 19. After deinterleaving, these bits are in positions 3, 8, and 13. Thus, the burst error has been separated by the depth of the interleaver (in our example, 5).

5.2.4 Concatenated Codes

In the attempt to create powerful codes with reasonable code rates by increasing the code length, the encoding and decoding complexity grow exponentially. One solution to this is to impose algebraic structure on the codes, as shown in section 5.3. Another technique is to concatenate two codes, as shown in Figure 5.4. An outer block coder first encodes message data. The block coder output then is interleaved.

The interleaver output is the input to an inner convolutional coder. Note that the nomenclature, inner and outer, is with respect to the channel, not the data. The inner code also is sometimes a block code, but all known DBS systems use a convolutional inner code. The concatenated decoder reverses the process shown in Figure 5.4, as shown in Figure 5.5.

The bitstream from the channel is first decoded by the inner convolutional decoder (Viterbi decoder). The Viterbi decoder output is deinterleaved and then decoded by the outer block decoder to create the message output.

Section 5.6 describes concatenated codes in more detail, and section 5.7 gives performance information for the important case where the outer block code is a Reed-Solomon code.

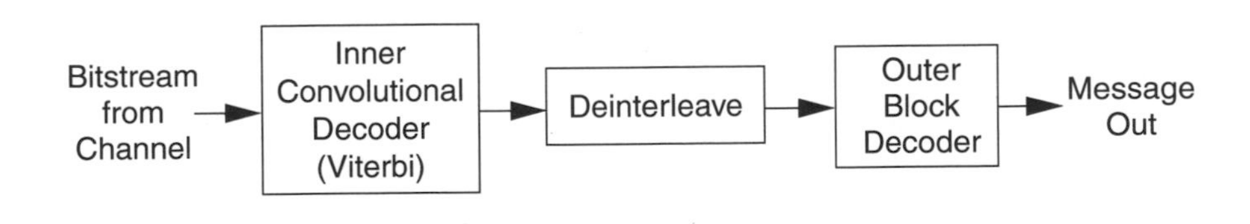

Figure 5.5 Concatenated Decoder

5.2.5 Code Performance

In what perhaps is the most famous equation in communication theory, Claude Shannon showed [Shannon48, 49] that if the information rate R in bits per second is less than the channel capacity C, written as

$$C = B\log_2\left[1+\frac{S}{N}\right] \tag{5.10}$$

where B is the channel bandwidth in hertz and S/N is the ratio of signal power to noise power, a code exists that can make the probability of error diminishingly small. Shannon did not tell us how to find such a code, only that it existed. Nevertheless, it provides a standard against which various coding schemes can be judged.

To determine the minimum E_b/N_o where error-free performance can result, we make the following substitutions into Equation (5.10): The signal power is the energy per bit E_b divided by the time per bit T_b; or $S = E_b/T_b$; the noise power N equals the noise power spectral density times the bandwidth, or $N = \mathbf{N_oB}$. Thus, (5.10) becomes

$$C = B\log_2\left[1+\frac{E_b/T_b}{N_oB}\right] \tag{5.11}$$

The optimum encoded signal is not restricted in bandwidth, so

$$C = \lim_{B\to\infty}\left\{B\log_2\left[1+\frac{E_b/T_b}{N_oB}\right]\right\}$$

Next, if we let $X = \frac{1}{B}$, then

$$C = \lim_{X\to 0}\left\{\frac{\log\left[1+\left(\frac{E_b}{N_oT_b}\right)X\right]}{X}\right\}$$

Using L'Hopital's rule, this becomes

$$\begin{aligned} C &= \lim_{X\to 0}\left\{\frac{1}{1+(E_b/N_oT_b)X}\left(\frac{E_b}{N_oT_b}\right)\log_2 e\right\} \\ &= E_b/(N_oT_b\ln 2) \end{aligned}$$

because $\log_2 e = \frac{1}{\ln 2}$. Since, in the limit $R_b \rightarrow C$,

$$C = \frac{1}{T_b} = \frac{E_b}{N_o T_b \ln 2}$$

$$\text{or} \quad \frac{E_b}{N_o} = \ln 2 = -1.59 \text{ dB}.$$

As will be shown in subsequent sections, none of the present-day coding systems approaches this Holy Grail. However, they do provide an absolute limit on how well we can do.

In section 3.7, it was noted that the E_b/N_o threshold for DBS systems was 5 dB. Thus, we are still almost 7 dB worse than the theoretical optimum, even when using a very powerful concatenated code.

Other Code Performance Issues

If the minimum distance of a code C is $d_{\min}$, any two distinct code vectors of C differ in at least $d_{\min}$ places. No error pattern of $d_{\min} - 1$ or fewer errors can change one code vector into another. It cannot detect all errors of $d_{\min}$ errors because at least two code vectors have a distance of $d_{\min}$ (by the definition of $d_{\min}$).

Even though a code with minimum distance $d_{\min}$ is capable of detecting all errors of $d_{\min} - 1$, it also is capable of detecting a much larger number of errors, $2^n - 2^k$ in fact. Among the $2^n - 1$ possible nonzero error patterns, there are $2^n - 1$ error patterns that are identical to the $2^n - 1$ code vectors. If any of these $2^n - 1$ error patterns occur, it maps one code vector into another. Thus, there are $2^n - 1$ undetectable errors. The total number of detectable errors is then the total number of error patterns minus the undetectable error patterns, or

$$(2^n - 1) - (2^k - 1) = 2^n - 2^k \tag{5.12}$$

If C is an (n, k) linear block code, A_i is the number of code vectors of weight i. The $\{A_i\}$ is called the weight distribution of the code. Because an undetectable error occurs only when the error pattern is identical to a nonzero code vector of C,

$$P_u(\text{E}) = \sum_{i=1}^{n} A_i p^i (1-p)^{n-i}, \tag{5.13}$$

where $P_u(\text{E})$ is the probability of an undetected error and p is the probability of an individual bit error.

For the (7,4) Hamming code, for example,

$$A_0 = 1, A_1 = A_2 = 0, A_3 = 7, A_y = 7, A_5 = A_6 = 0, \text{and } A_7 = 1$$
$$\therefore P_u(\mathrm{E}) = 7 * p^3(1 - p)^4 + 7 * p^2(1 - p)^3 + p^7$$

If $p = 10^{-2}$, for example, $P_u(\mathrm{E}) \cong 7 * 10^{-6}$.

5.3 Block Codes: Cyclic Codes

The decoding of block codes can become quite complex. By utilizing a code with algebraic structure, decoding can be simplified. One class of such codes is cyclic codes. Cyclic codes have the property that all shifts of a code word are also code words.

5.3.1 Binary Cyclic Codes

Binary cyclic codes are cyclic codes in which all the elements are either a '1' or a '0'.

Mathematical Preliminaries

As previously noted, a field with a finite number of elements is called a finite field. Such fields are called Galois fields (GF) after the French mathematician who discovered them. These fields are designated as GF(q) where q is a prime number or its extension q^m. The case GF(2) is an important special case, where addition and subtraction are modulo-2 as previously defined.

Polynomials can be defined over GF(2). Such polynomials can be written as

$$f(x) = f_0 + f_1x + f_2x^2 + \dots f_\eta x^\eta \tag{5.14}$$

where the f_i are 0 or 1. These polynomials can be added, subtracted, multiplied, and divided. When $f(x)$ is divided by $g(x)$, we obtain a pair of unique polynomials $q(x)$, the quotient, and $r(x)$, the remainder. As an example, divide $f(x) = 1 + x + x^4 + x^5 + x^6$ by $g(x) = 1 + x + x^3$.

$$\begin{array}{r l}
 & x^3 + x^2 \\
x^3 + x + 1 & \overline{)\, x^6 + x^5 + x^4 \qquad\qquad + x + 1} \\
 & \underline{x^6 \qquad + x^4 + x^3} \\
 & \quad x^5 \qquad\quad x^3 \quad\; x \quad 1 \\
 & \quad \underline{x^5 \qquad\quad x^3 \;\; x^2} \\
 & \qquad\qquad\quad x^2 \; x \quad 1
\end{array} \tag{5.15}$$

Thus, the quotient $q = x^3 + x^2$ and the remainder $r = x^2 + x + 1$.

As with normal polynomial arithmetic, if $r(x) = 0$, we say that $f(x)$ is divisible by $g(x)$ and that $g(x)$ is a factor of $f(x)$. For real numbers, if a is a root of a polynomial $f(x)$, then $f(x)$ is divisible by $(x - a)$. This is still true over GF(2).

Example 5.3 According to Lin and Costello [Lin+83], let

$$f(x) = 1 + x^2 + x^3 + x^4$$

$$f(1) = 1 + 1^2 + 1^3 + 1^4 = 0$$

so $f(x)$ is divisible by $x + 1$. This can be verified by carrying out the division, which yields

$$q = x^3 + x + 1$$

It is easy to see that any polynomial $f(x)$ over GF(2) that has an even number of terms has 1 as a root and $x + 1$ as a divisor.

A polynomial $P(x)$ over GF(2) of degree m is said to be irreducible over GF(2) if $P(x)$ is not divisible by any polynomial over GF(2) that has degree greater than 0 and less than m. It has been shown that for any $m \geq 1$ there exists an irreducible polynomial of degree m. It also can be shown that any irreducible polynomial over GF(2) of degree m divides $x^{2^m-1}+1$.

An irreducible polynomial $P(x)$ of degree m is said to be primitive if the smallest positive integer n for which $P(x)$ divides $x^n + 1$ is $n = 2^m - 1$. Finally, it can be shown that, if $f(x)$ is a polynomial over GF(2), then

$$[f(x)]^{2^i} = f\left(x^{2^i}\right) \tag{5.16}$$

If an n-tuple,

$$v = (v_0, v_1, \ldots v_{n-1})$$

is rotated to the right one place, we obtain

$$v^{(1)} = (v_{n-1}, v_0, v_1 \ldots v_{n-2})$$

If it is rotated i places to the right, the resultant n-tuple is

$$v^{(i)} = (v_{n-i}, v_{n-i+1}, \ldots v_{n-i-1}).$$

An (n, k) code C is called cyclic if every right rotation of a code vector in C is itself a code vector in C. Cyclic codes form a very important subclass of

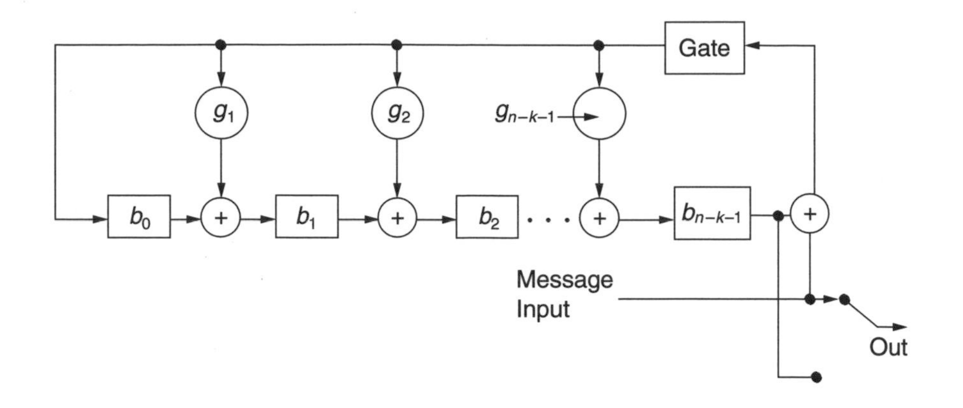

Figure 5.6 Encoding for (n, k) Cyclic Codes From [Lin+83]. Copyright © Prentice-Hall, 1983. Reprinted by permission of Prentice-Hall, Inc., Upper Saddle River, NJ.

linear block codes. To develop the properties of cyclic codes, we treat the code vectors as polynomials

$$v(x) = v_0 + v_1 x + v_2 x^2 + \dots + v_{n-1} x^{n-1} \tag{5.17}$$

If $v_{n-1} \neq 0$, the degree of $v(x)$ is $n - 1$.

Figure 5.6 shows an encoding circuit for an (n, k) cyclic code. First with the gate on, the k are shifted into the circuit and the communication channel. The gate is then opened and the $n - k$ parity digits are shifted into the communication channel.

If

$$r(x) = r_0 + r_1 x + \dots r_{n-1} x^{n-1}$$

is the polynomial for a received vector, then dividing $r(x)$ by $g(x)$ yields

$$r(x) = a(x)g(x) + s(x) \tag{5.18}$$

Figure 5.7 shows a circuit for creating the syndrome. The shift-register state starts at all 0s.

After all of the n bits of $r(x)$ are shifted through the circuit, the contents of the shift register contain the syndrome. Once the syndrome has been calculated, it is straightforward to calculate the correct code vector.

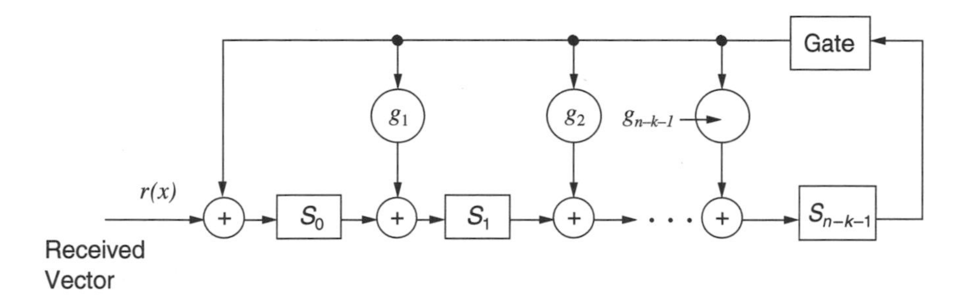

Figure 5.7 Syndrome Calculating Circuit From [Lin+83]. Copyright © Prentice-Hall, 1983. Reprinted by permission of Prentice-Hall, Inc., Upper Saddle River, NJ.

5.3.2 BCH Codes

For any positive integers $m \geq 3$, and $t < 2^{m-1}$, a BCH code is defined by the following:

Block length: $b = 2^m - 1$
Number of parity check digits: $n - k \leq m * t$
Minimum distance: $d_{\min} \leq 2t + 1$

This is a t-error correcting BCH code. Appendix D contains a detailed discussion of BCH codes.

5.4 Reed-Solomon Codes

In addition to the binary codes, such as the BCH codes discussed in section 5.3.2, there are nonbinary block codes. If p is a prime and $q = p^m$ where m is a positive integer, there are codes with symbols from the Galois field GF(q). These codes are called q-ary codes. The preceding techniques for encoding and decoding binary codes generalize in a straightforward way to nonbinary codes.

For any choice of positive integers s and t there exists a q-ary BCH code of length $q^s - 1$, which is capable of correcting any combination of t or fewer errors and requires no more than $2 * s * t$ parity digits. The case $s - 1$ is the most important and these codes are called Reed-Solomon codes after their discoverers [Reed+60]. Also, most of the important applications have $p = 2$, so $q = 2^m$. For all known DBS systems, the symbols are bytes, so $m = 8$. Thus, $q = 2^8$.

A t-error correcting Reed-Solomon code is thus defined by

Block length: $n = q - 1$

Number of parity check digits: $n - k = 2t$

Minimum distance: $d_{min} = 2t + 1$

If we let $q = 2^8 = 256$, then $n = 255$. Further, if we let $t = 8$ so that eight bytes worth of errors can be corrected, a (255, 239) code results. This has been a NASA-standard Reed-Solomon code for a number of years. We can form a code with $n' = n - 1$ and $k' = k - 1$. This code is called a shortened Reed-Solomon code.

The two major DBS implementations (DIRECTV and DVB) use shortened Reed-Solomon codes. DIRECTV uses a (146, 130) shortened Reed-Solomon code ($1 = 109$) and DVB uses a (204, 188) shortened Reed-Solomon code ($1 = 51$), both of which can correct eight bytes of errors per information packet.

Let α be a primitive element in GF(2^8). The generator polynomial of a primitive t-error correcting Reed-Solomon code of length $2^m - 1$ is

$$\begin{aligned} g(x) &= (x+\alpha)(x+\alpha^2) \ldots (x+\alpha^{2t}) \\ &= g_0 + g_1 x + \ldots g_{2t-1}x^{2t-1} + x^{2t} \end{aligned} \tag{5.19}$$

The code generated by $g(x)$ is an $(n, n - 2t)$ cyclic code that consists of those polynomials of degree $n - 1$ or less with coefficients that are multiples of $g(x)$.

Encoding of this code is similar to the binary case. Let

$$a(x) = a_0 + a_1 x + \ldots a_{k-1}x^{k-1} \tag{5.20}$$

be the message to be encoded where $k = n - 2t$. In systematic form, the $2t$ parity-check digits are the coefficients of

$$b(x) = b_0 + b_1 x + \ldots + b_{2t-1}x^{2t-1} \tag{5.21}$$

resulting from dividing $x^{2t}\alpha(x)$ by $g(x)$. It can be shown that the minimum distance of a Reed-Solomon code is exactly $2t + 1$.

Let

$$v(x) = v_0 + v_1 x + \ldots v_{n-1}x^{n-1}$$

be the transmitted code vector and

$$r(x) = r_0 + r_1 x + \ldots r_{n-1}x^{n-1}$$

be the corresponding received vector. The error pattern then is

$$e(x) = e_0 + e_1 x + \dots e_{n-1} x^{n-1} \tag{5.22}$$

where $e_i = r_i - v_i$ is a symbol from GF(2^8).

The same three steps for decoding the BCH codes are used to decode the Reed-Solomon codes. The real generalization here is that if there is a symbol error, both the location of the error and the correction value must be determined. In binary codes, the value is easy because the bit is changed if an error has been made. For nonbinary codes, the symbol also must be corrected. Entries in the equivalent of Table D.1 in Appendix D are of the form $1 + \alpha^i x^j$. If $Z(x)$ is defined as

$$\begin{aligned} Z(x) = {} & 1 + (s_1 + \sigma_1)x + (s_2 + \sigma_1 s_1 + \sigma_2)x^2 + \dots \\ & + (S_v + \sigma_1 S_{v-1} + \sigma_2 S_{v-2} + \dots + \sigma_v)x^v \end{aligned} \tag{5.23}$$

then the error value at location $\beta_l = \alpha^{jl}$ is given by

$$e_{j_l} = \frac{Z(\beta_l^{-1})}{\prod_{\substack{i=1 \\ i \neq l}}^{v} (1 + \beta_i \beta_l^{-1})} \tag{5.24}$$

Weight of Reed-Solomon Codes The number of code vectors of weight j is

$$A_j = \binom{2^m - 1}{j} * \sum_{i=0}^{j-2*t-1} (-1)^i * \left(i^j\right) * \left(2^{m*(j-2*t-i)}\right) \tag{5.25}$$

5.5 Interleaver

In section 5.2, interleavers were introduced. Some DBS systems use more elaborate interleavers. This section, however, will discuss what are called "Ramsey Interleavers" [Ramsey70].

An (n_2, n_1) interleaver reorders a sequence so that no contiguous sequence of n_2 symbols in the reordered sequence contains any symbols that are separated by fewer than n_1 symbols in the original ordering.

Specifically, the DVB standard specifies a Ramsey Type III interleaver. Figure 5.8 is a block diagram of a general Ramsey Type III interleaver. The Type III interleaver consists of an $[(n_2 - 1)(n_1 + 1) + 1]$ stage shift register

Tap Number	n_2-1	n_2-2	$k-1$	1	0
Stage Number	$(n_2-1)(n_1+1)$	$(n_2-2)(n_1+1)$	$(k-1)(n_1+1)$	n_1+1	0

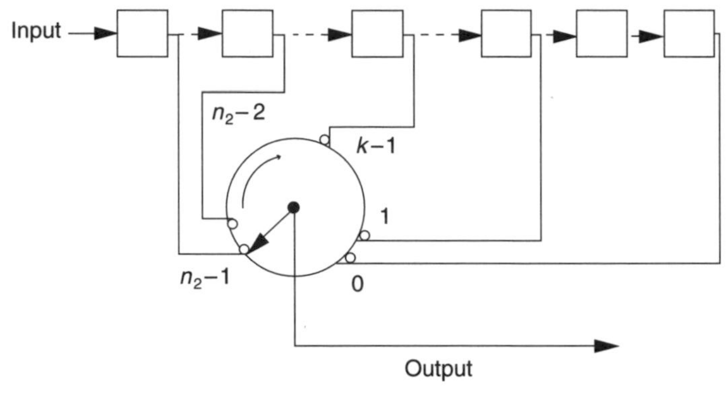

Figure 5.8 Ramsey Type III Interleaver From [Ramsey70]. Copyright © IEEE, 1970. Reprinted by permission of IEEE, Piscataway, NJ.

with taps at the outermost stages and every $(n_1 + 1)$-th intermediate stage. It employs an n_2 position commutator that cyclically samples the n_2 taps in the same order as their distance from the input of the shift register.

To demonstrate the operation of the Type III interleaver, consider the example where $n_1 = 3$ and $n_2 = 2$, which are relatively prime. The quantity $[(n_2 - 1)(n_1 + 1) + 1]$ becomes $[(1)(4) + 1] = 5$, and there are $n_2 = 2$ taps.

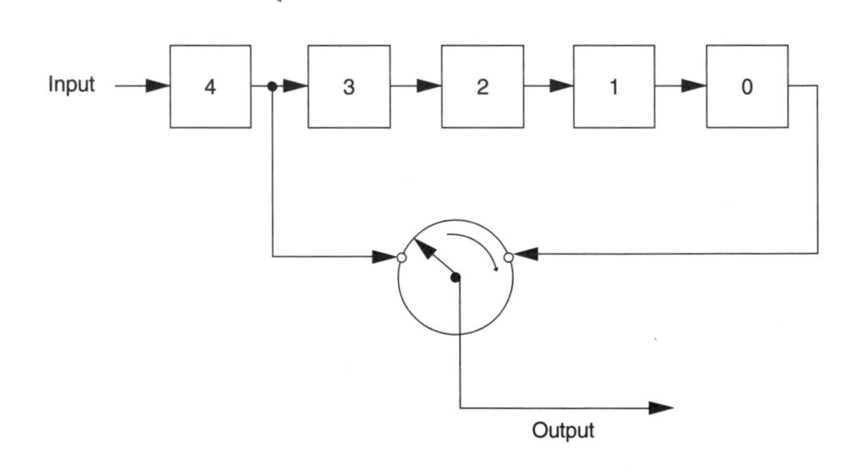

Figure 5.9 Block Diagram of Example (3, 2) Interleaver From [Ramsey70]. Copyright © IEEE, 1970. Reprinted by permission of IEEE, Piscataway, NJ.

Figure 5.9 is the block diagram for this example. Table 5.2 shows a sequence of input symbols. The entries show where these input symbols are in the shift register.

Figure 5.10 is the block diagram for the deinterleaver for the Type III interleaver (Ramsey used the term *unscrambler* in his original paper, but I believe that *deinterleaver* is a better word). It should be noted that the Type III deinterleaver is an interleaver in its own right and is called a Type IV interleaver. If the commutator tap is not at the location corresponding to that ⊕ then the output of the prior stage is shifted into the next stage. If the commutator is at that location, the input is shifted into the next stage.

Table 5.2 Example (3, 2) Interleaver Process

Stage 4	Stage 3	Stage 2	Stage 1	Stage 0	Output
0	0	0	0	0	
9	0	0	0	0	9
8	9	0	0	0	0
7	8	9	0	0	7
6	7	8	9	0	0
5	6	7	8	9	5
4	5	6	7	8	8
3	4	5	6	7	3
2	3	4	5	6	6
1	2	3	4	5	1
0	1	2	3	4	4
−1	0	1	2	3	−1
−2	−1	0	1	2	2
−3	−2	−1	0	1	−3
−4	−3	−2	−1	0	0
−5	−4	−3	−2	−1	−5
−6	−5	−4	−3	−2	−2
−7	−6	−5	−4	−3	−7

Tap Number	n_2-1	n_2-2	$k-1$		0
Stage Number	$(n_2-1)(n_1+1)$	$(n_2-2)(n_1+1)$	$(k-1)(n_1+1)$	1	0

Figure 5.10 Ramsey III Deinterleaver From [Ramsey70]. Copyright © IEEE, 1970. Reprinted by permission of IEEE, Piscataway, NJ.

Continuing with the (3, 2) example in Figure 5.11, the interleaver output was

–7 –2 –5 0 –3 2 –1 4 1 6 3 8 5 0 7 0 9

The deinterleaver proceeds according to Table 5.3. The DVB standard requires an (**N**, **M**) Ramsey Type III interleaver.

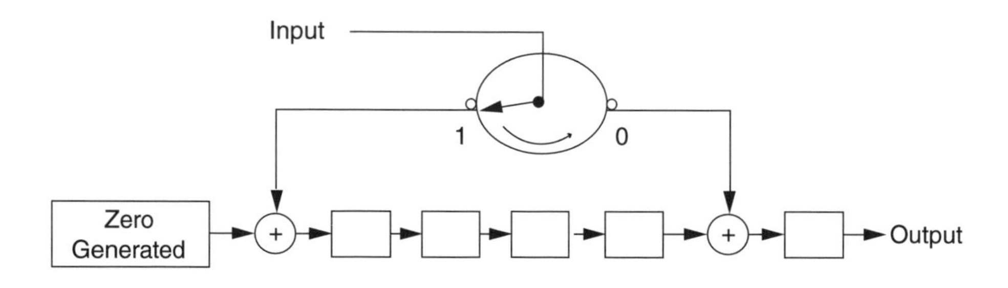

Figure 5.11 Deinterleaver (3, 2) Example From [Ramsey70]. Copyright © IEEE, 1970. Reprinted by permission of IEEE, Piscataway, NJ.

Table 5.3 Ramsey Deinterleaver

4	3	2	1	0	Output
9	0	0	0	0	x
0	9	0	0	0	x
7	0	9	0	0	x
0	7	0	9	0	x
5	6	7	0	9	x
0	5	6	7	8	9
3	0	5	0	7	8
0	3	0	5	6	7
1	0	3	0	5	6
0	1	0	3	4	5
–1	0	1	0	3	4
0	–1	0	1	2	3
–3	0	–1	0	1	2
0	–3	0	–1	0	1
–5	0	–3	0	–1	0
0	–5	0	–3	–2	–1
–7	0	–5	0	–3	–2
0	0	0	–5	–4	–3
x	x	0	–8	–5	–4
x	x	x	0	–6	–5
x	x	x	x	x	–6

5.6 Convolutional Codes/Viterbi Decoding

Convolutional codes were introduced in section 5.2.2. In this section, we revisit convolutional codes with particular emphasis on the Viterbi algorithm for decoding convolutional codes and specific codes used in DBS systems.

Convolutional codes are generated by shift registers, modulo-2 adders (exclusive OR circuits), and multiplexers. These codes are described as (n, k, m)

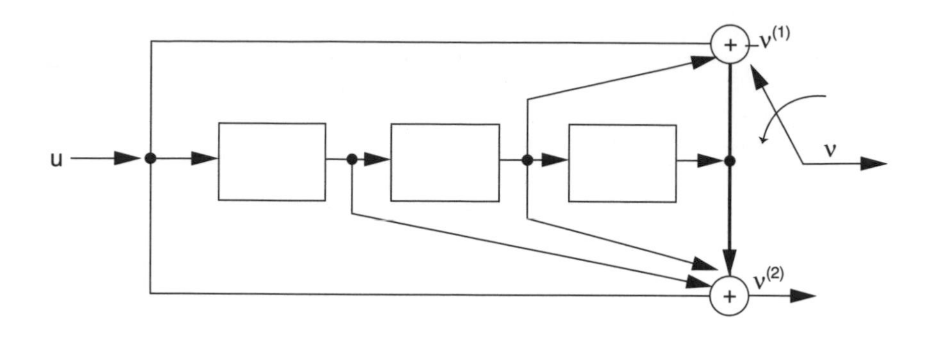

Figure 5.12 (2, 1, 3) Convolutional Encoder From [Lin+83]. Copyright © Prentice-Hall, 1983. Reprinted by permission of Prentice-Hall, Inc., Upper Saddle River, NJ.

convolutional codes, where n and k have the same relationship as in block codes. The parameter m is a new parameter called the constraint length. If $k = 1$, that is, the code is $(n, 1, m)$, there is one shift register and m is the length of that shift register plus one. However, if $k > 1$, there may be multiple shift registers with varying lengths. The modern viewpoint is to use a single shift register and what are called punctured codes to generate rational fraction-code-rate convolutional codes. The code rate of a convolutional code, as with a block code, is k/n.

Figures 5.12, 5.13, and 5.14 show convolutional encoders with small values of m. In addition, Figures 5.13 and 5.14 show how $r = k/n$ rate convolutional codes can be constructed when $k \geq 1$. As noted, these codes also can be generated from a $k = 1$ encoder by a technique called puncturing [Viterbi95].

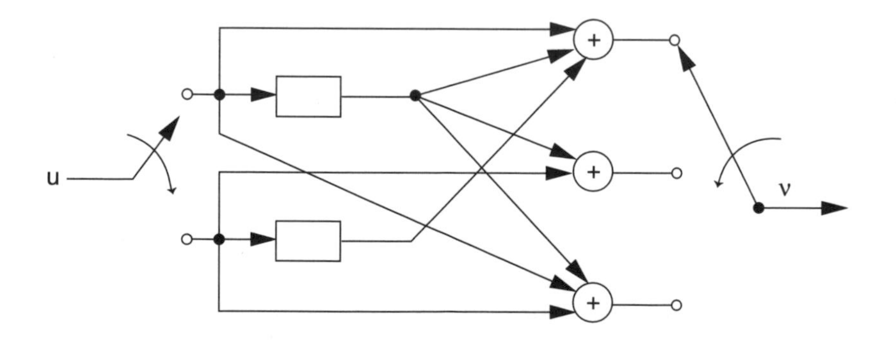

Figure 5.13 (3, 2, 1) Convolutional Encoder From [Lin+83]. Copyright © Prentice-Hall, 1983. Reprinted by permission of Prentice-Hall, Inc., Upper Saddle River, NJ.

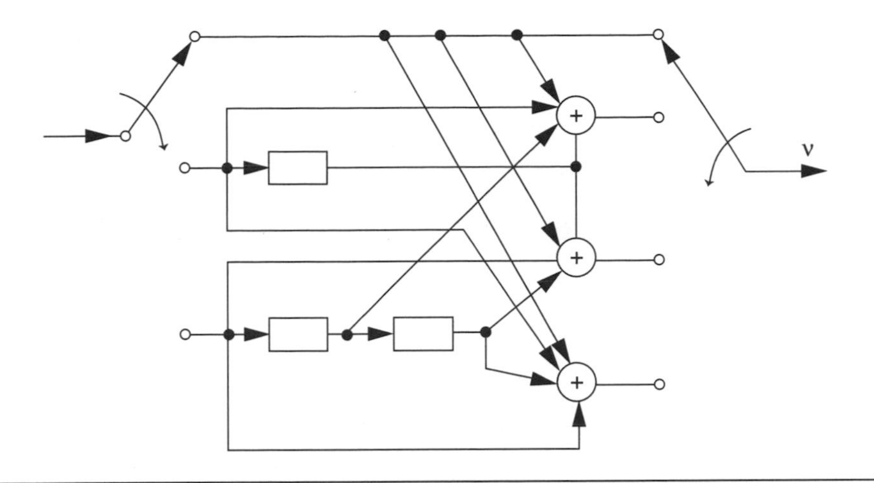

Figure 5.14 (4, 3, 2) Convolutional Encoder From [Lin+83]. Copyright © Prentice-Hall, 1983. Reprinted by permission of Prentice-Hall, Inc., Upper Saddle River, NJ.

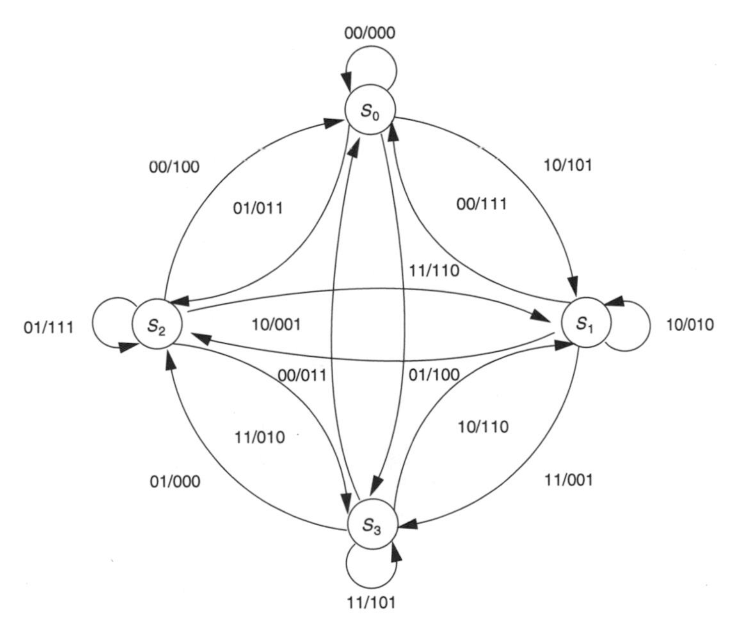

Figure 5.15 State Diagram for (3, 2, 1) Convolutional Encoder From [Lin+83]. Copyright © Prentice-Hall, 1983. Reprinted by permission of Prentice-Hall, Inc., Upper Saddle River, NJ.

A convolutional encoder is a finite-state machine and consequently can be described by a state diagram. Figure 5.15 shows the state diagram for the Figure 5.13 encoder; circles represent the encoder states. There are four such circles representing the four possible states of the two storage elements.

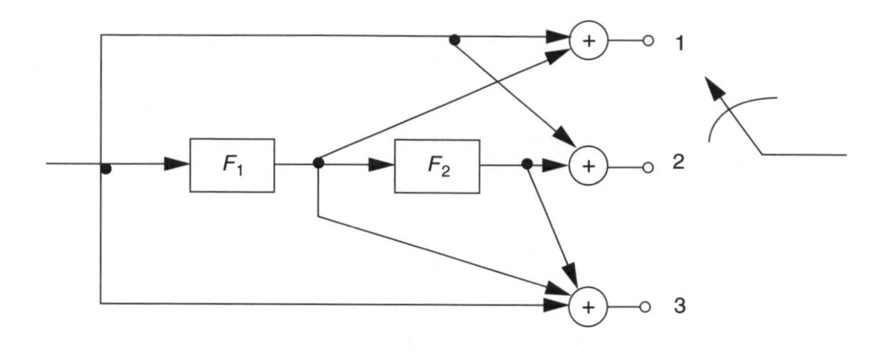

Figure 5.16 (3, 1, 2) Convolutional Encoder From [Lin+83]. Copyright © Prentice-Hall, 1983. Reprinted by permission of Prentice-Hall, Inc., Upper Saddle River, NJ.

The labeling convention for the branches is input/output. Consider the self-loop on S_2: If 01 is the input, the encoder stays in state 2 and outputs 111. There are 2^k branches leaving each node. Figure 5.16 is the encoder for a (3, 1, 2) convolutional code and Figure 5.17 is its state diagram.

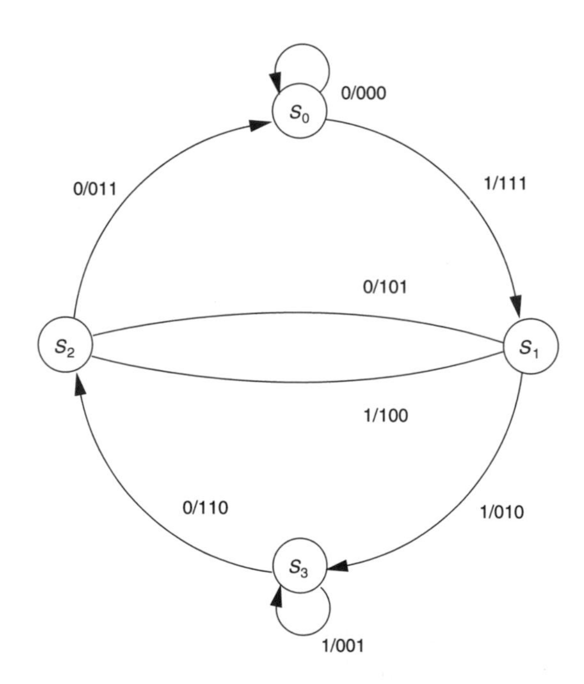

Figure 5.17 State Diagram for (3, 1, 2) Convolutional Encoder From [Lin+83]. Copyright © Prentice-Hall, 1983. Reprinted by permission of Prentice-Hall, Inc., Upper Saddle River, NJ.

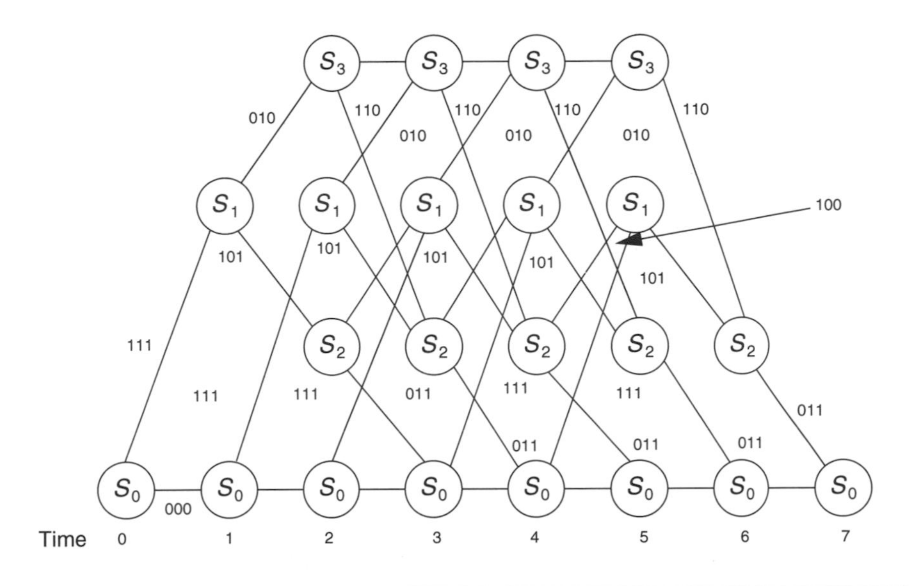

Figure 5.18 Trellis Diagram for (3, 1, 2) Convolutional Code From [Lin+83]. Copyright © Prentice-Hall, 1983. Reprinted by permission of Prentice-Hall, Inc., Upper Saddle River, NJ.

5.6.1 Viterbi Decoding

In 1967, Andrew Viterbi introduced the algorithm that bears his name. Subsequently, other applications have been found for the algorithm [Forney73].

To understand the Viterbi algorithm, it is convenient to introduce the concept of a trellis diagram. The trellis diagram can be considered as an expansion of the state diagram at each unit of time. If CL represents the length of an input sequence, then Figure 5.18 shows the trellis diagram for the (3, 1, 2) encoder shown in Figure 5.15 with CL = 2.

In Figure 5.18, the branch labeling indicates only the output. The upper branch from a state represents a '1' input and the lower branch is a '0' input. The first m time units correspond to the departure from state 0 and the last m units correspond to the return to S_0.

If the vector $\mathbf{r}$ is received when the code word $\mathbf{v}$ is transmitted, then the maximum likelihood decoder chooses $\hat{v}$ as the code word that maximizes the log likelihood function $\log\left[P[\mathbf{r} \mid \mathbf{v}]\right]$. For a discrete memoryless channel (DMC),

$$\mathrm{P}[\mathbf{r}|\mathbf{v}] = \prod_{i=0}^{L+m-1} \mathrm{P}[\mathbf{r}_i|\mathbf{v}_i] = \prod_{i=0}^{N-1} \mathrm{P}[\mathbf{r}_i|\mathbf{v}_i] \tag{5.26}$$

Thus,

$$\log\left[P[\mathbf{r}|\mathbf{v}]\right] = \sum_{i=0}^{L+m-1} \log\left[P[\mathbf{r}_i|\mathbf{v}_i]\right] = \sum_{i=0}^{N-1} \log\left[P[\mathbf{r}_i|\mathbf{v}_i]\right] \tag{5.27}$$

The log likelihood function log $[P[\mathbf{r}|\mathbf{v}]]$ is called the metric associated with path v and is noted by $M[\mathbf{r}|\mathbf{v}]$. The individual terms log $[P[\mathbf{r}_i|\mathbf{v}_i]]$ are called r; the branch metrics, and the log $[P[\mathrm{r}_i|\mathrm{v}_i]]$, are called the bit metrics. The path metric, $M[\mathbf{r}|\mathbf{v}]$, can then be written as

$$M[\mathbf{r}|\mathbf{v}] = \sum_{i=0}^{L+m-1} M[\mathbf{r}_i|\mathbf{v}_i] = \sum_{i=0}^{N-1} M[\mathrm{r}_i|\mathrm{v}_i] \tag{5.28}$$

A partial path metric for the first j branches can be written as

$$M\left([\mathbf{r}|\mathbf{v}]_j\right) = \sum_{i=0}^{j-1} M[\mathbf{r}_i|\mathbf{v}_i] \tag{5.29}$$

The Viterbi algorithm finds the path through the trellis with the largest metric. It processes **r** in an iterative manner. At each step it selects the path entering the state with the highest metric (the survivor) and discards all others.

5.6.2 Viterbi Algorithm

Step 1. Beginning at time unit $j = m$, compute the partial metric for the single path entering each state. Store the path (the survivor) and its metric for each state.
Step 2. Increase j by 1. Compute the partial metric for all the paths entering a state by adding the branch metric entering that state to the metric of the connecting survivor at the preceding time unit. For each state, store the path with the largest metric (the survivor) together with its metric and eliminate all other paths.
Step 3. If $j < L + m$, repeat Step 2. Otherwise, stop.

The individual entities, log $[P[r_i|v_i]]$, take on negative fractional values, with the largest being the least negative. It is more convenient to use positive integers. Because the path that maximizes

$$\sum_{i=0}^{N-1} \log\left[P[r_i|v_i]\right]$$

also maximizes

$$\sum_{i=0}^{N-1} C_2\left[\log\left[P[r_i|v_i]\right] + C_1\right]$$

C_1 and C_2 can be chosen so that the metrics are positive integers.

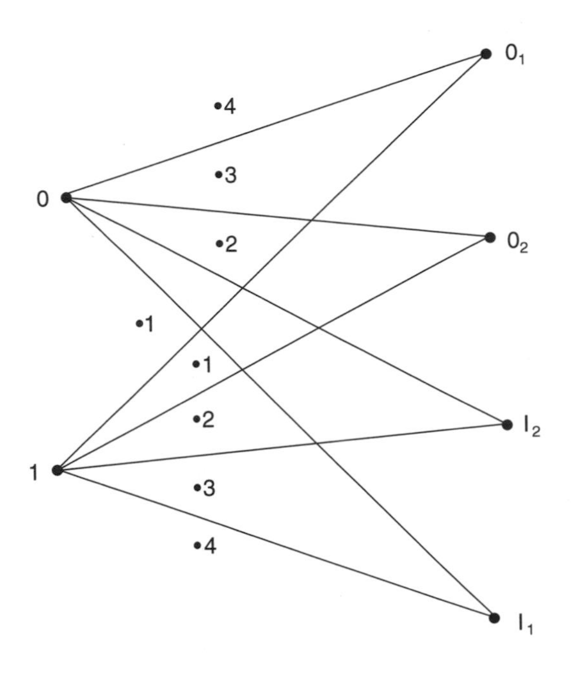

Figure 5.19 DMC with Four Outputs From [Lin+83]. Copyright © Prentice-Hall, 1983. Reprinted by permission of Prentice-Hall, Inc., Upper Saddle River, NJ.

If C_1 is chosen to make the smallest metric zero, C_2 can be chosen so that all of the other metrics can be approximated by integers. As an example, consider the DMC (see Figure 5.19) with binary inputs and quatenary-received sequence defined by the transition probability matrix. Figure 5.20 shows the $\log[P[r_i | v_i]]$ values.

V_i \ λ_I	0_1	0_2	I_2	I_1
0	–.4	–.52	–.7	–1
1	–1	–.7	–.52	–.4

Figure 5.20 Branch Metrics Example

V_i \ λ_I	0_1	0_2	I_2	I_1
0	10	8	5	0
1	0	5	8	10

Figure 5.21 Example with Adjusted Metric

Choosing $C_1 = 1$, the other metrics become

$$.6 \quad .48 \quad .3.$$

This means C_2 can be chosen so that the metrics are

$$10 \quad 8 \quad 5.$$

Figure 5.20 can then be written as shown in Figure 5.21.

Figure 5.22 shows the (3, 1, 2) trellis diagram for this example. The first branch received is $1_1\, 1_2\, 0_1$. If an input 1 with 111 transmitted was sent, the metric is

	1_1	1_2	0_1	
1	10	8	0	= 18
0	0	5	10	= 15

These are the labels on the time 1, state 1, and state 0. At time 2, $1_1\, 1_1\, 0_2$ is received as shown in Table 5.4.

At time 3, $1_1\, 1_1\, 0_1$ is received. Entering State 3 are 001 from State 3 and 010 from State 1 (see Table 5.5). The survivor is the branch from S_1, and the branch from S_3 is eliminated (see X on branch in Figure 5.23).

The process continues with the losing branch receiving an X. Finally, when we get to 0 at time 7, only one path through the trellis is not blocked by an X. This path is shown in bold in Figure 5.22.

In the special case of a binary symmetric channel (BSC), each received bit is declared a 1 or a 0 (rather than the four possibilities of the previous example). It can be shown that maximizing the branch metric is equivalent to minimizing the Hamming distance. Repeating the (3, 1, 2) example for the BSC, Figure 5.23 shows the implementation of the Viterbi algorithm in this case.

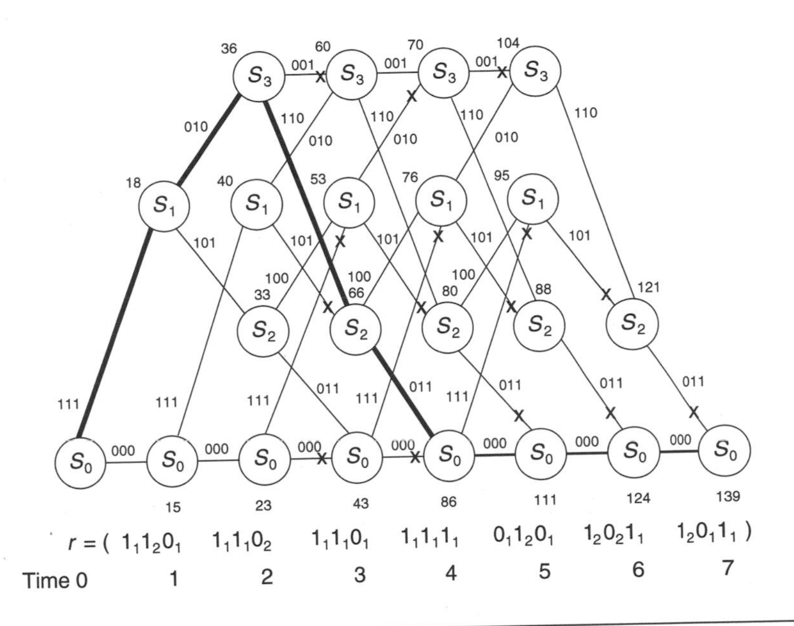

Figure 5.22 Trellis Diagram for (3, 1, 2) Example From [Lin+83]. Copyright © Prentice-Hall, 1983. Reprinted by permission of Prentice-Hall, Inc., Upper Saddle River, NJ.

Table 5.4 Viterbi Algorithm Example—Time 2

1_1	1_1	0_2				Prior	Total	State
0	1	0	0+	10+	8 = 18	18	36	S_3
1	0	1	10+	0+	5 = 15	18	33	S_2
1	1	1	10+	10+	5 = 25	15	40	S_1
0	0	0	0+	0+	8 = 8	15	23	S_0

Table 5.5 Viterbi Algorithm Example—Time 3

1_1	1_1	0_1				Prior	Total
0	0	1	0+	0+	10 = 10	S_3 = 36	46
0	1	0	0+	10+	10 = 20	S_1 = 40	60

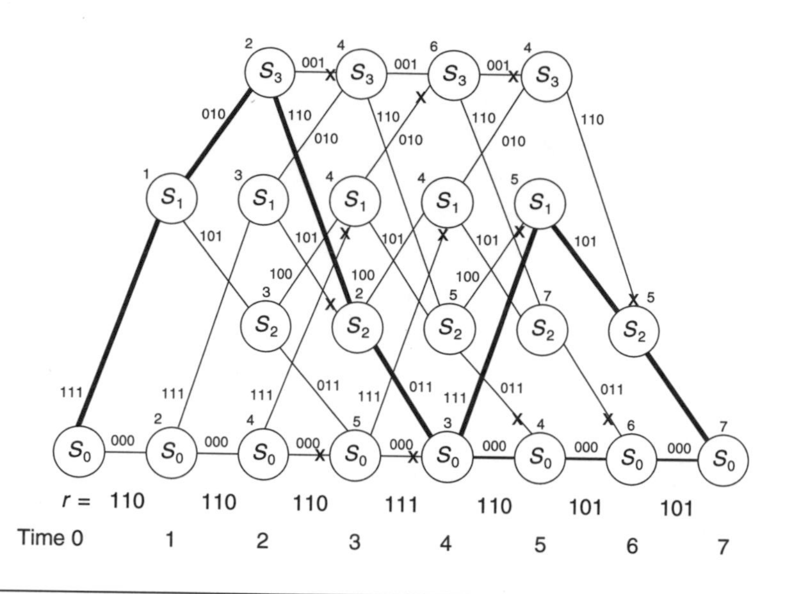

Figure 5.23 Trellis Diagram for BSC From [Lin+83]. Copyright © Prentice-Hall, 1983. Reprinted by permission of Prentice-Hall, Inc., Upper Saddle River, NJ.

Note the excursion off to State 1 at time 5. This represents a bit error in the final decoded output.

The last two examples illustrate a critical point: If the demodulator makes soft decisions, improved performance results. In the case of infinite-quantization soft decisions, the advantage over hard decisions is 2 dB. In the practical case of eight-level quantization, there is about a 0.2-dB loss compared to infinite quantization. Thus, an eight-level quantization has a 1.75-dB advantage over hard decisions.

5.7 Performance

The key to any coding system is its performance. Code performance is usually presented by plotting the probability of bit error P_b versus E_b/N_o which is the energy per bit divided by the noise power spectral density.

5.7.1 Inner (Convolutional) Code

With Viterbi decoding, the performance of a convolutional code depends on the distance properties of the code. The minimum free distance is

$$d_{free} \Delta \min[d[v', v'']: u' \geq u'']$$

where v' and v'' are the code words corresponding to inputs u' and u'', respectively. Hence, d_{free} is the minimum distance between any two code words in the code.

$$d_{free} = \min[w(\mathbf{v}' + \mathbf{v}'') : u' \neq u''] = \min[w(\mathbf{v}) : u \neq 0]$$

Notice that d_{free} has been calculated for a number of codes.

Table 5.6 shows the key parameters for a number of convolutional codes of practical interest for DBS [Proakis89, Daut+82].

Analysis of the performance of a particular code treats the convolutional encoder as a finite-state machine. The state equations can be written as

$$\mathbf{S} = [\mathbf{A}]\mathbf{S} + \mathbf{b} \tag{5.30}$$

Vector **S** represents the nonzero states and **A** represents the state transition matrix. Vector **b** represents the entry into the higher-order states from state 0. The transfer function of the finite-state machine then can be written as

$$\mathbf{T}(\omega, \mathbf{I}) = \mathbf{c} \cdot \mathbf{S} \tag{5.31}$$

From (5.30), we get $\mathbf{S}[\mathbf{I} - \mathbf{A}] = \mathbf{b}$ or,

$$\mathbf{S} = [\mathbf{I} - \mathbf{A}]^{-1} * \mathbf{b} \tag{5.32}$$

Thus,

$$\mathbf{T}[\omega, \mathbf{I}] = \mathbf{c} * [\mathbf{I} - \mathbf{A}]^{-1} * \mathbf{b} \tag{5.33}$$

It can be shown [Viterbi95, Wilson96, Clark+81] that the probability of a bit error is

$$P_b \leq \frac{1}{k\left(2\pi d_f r E_b / N_o\right)^{1/2}} \left. \frac{\partial T[\omega, I]}{\partial I} \right|_{\substack{I=1 \\ \omega = e^{-rE_b/N_0}}} \tag{5.34}$$

Several authors [Viterbi95, Wilson96] note that $[\mathrm{I} - \mathrm{A}]^{-1}$ can be evaluated as

$$[\mathrm{I} - \mathrm{A}]^{-1} = \mathrm{I} + \mathrm{A}[\omega] + \mathrm{A}^2[\omega] + \ldots \tag{5.35}$$

and

$$\left. \frac{\partial T[\omega, I]}{\partial I} \right|_{I=1}$$

Table 5.6 Key Parameters

Rate	Constraint Length	Generators (Octal)	d_{free}
1/2	4	15, 17	6
1/2	5	23, 35	7
1/2	6	53, 75	8
1/2	7	133, 171	10
1/2	9	561, 753	12
2/3	7	177, 055, 112	6
3/4	7	127, 045, 106, 172	5
5/6	7	025, 071, 123, 046, 111, 175	4
6/7	7	003, 005, 010, 021, 041, 101, 176	3
7/8	7	generated Cr by punctured code	3

can be evaluated as

$$\frac{T[\omega,\ 1+\delta]-T[\omega,\ 1]}{\delta}$$

for small δ. However, in the performance curves given later in this section, the software program Mathematica was used to form the symbolic matrix inverse $[I - A]^{-1}$ and the symbolic derivative

$$\frac{\partial T[\omega,\ I]}{\partial I} \tag{5.36}$$

This expression then was evaluated at $I = 1$, $\omega = e^{-rE_b/N_o}$.

In any event, the result is a big mess, even for $k = 1$. For example, the matrix inverse for the (2, 1, 5) code took 17 pages to list.

There remains the problem of constructing **A**, **b**, and **c**. With CL as the constraint length, **b** and **c** are $2^{CL-1} - 1$ component vectors and **A** is a $2^{CL-1} - 1$ by $2^{CL-1} - 1$ matrix. The nonzero entries correspond to possible state transitions. The entries are a product of powers of W and I. The powers of I correspond to the 1s in the input for that branch and the powers of ω to the number of 1s in the output when that branch is traversed.

5.7.2 Rate 1/2 Codes

Determining the performance of rate 1/2 codes helps explain the procedure so that the analysis of more complicated codes can be understood and is important in its own right as one of the DVB options, a (2, 1, 7) code. Also all k/n codes can be generated from a rate 1/2 code as a punctured code.

Forming **b** and **c** is easy for rate = 1/2 codes. The first component of **b** is $I\omega^2$ because to leave state 0, a 1 input is required and the generator functions for all CL generate 11 for this transition. All other components of **b** are 0. All components of **c** are 0 except the middle term because the state with the MSB equals 1 and all other elements are 0.

To form **A**, note that the first row is unique. State 1 can be reached only from the nonzero state 2^{CL-2} with a 1 input. Thus, there is only one nonzero element, in the middle element, and the contribution from the input is I. All of the remaining rows are formed with two nonzero elements per row according to the following rules:

Even Row I, 1st Entry

location: $I/2$

contribution from input: 1

Odd Row I, 1st Entry

location: $(I - 1)/2$

contribution from input: I

Even Row I, 2nd Entry

location: $(I/2) + 2^{CL-2}$

contribution from input: 1

Odd Row I, 2nd Entry

location: $((I - 1)/2) + 2^{CL-2}$

contribution from input: I

Next, the contribution to the nonzero elements of **A** from the output must be determined. The outputs for each state transition are determined by generator functions, as shown in Table 5.6. Note that the generators are the connection information from the shift register stages to the exclusive OR circuits that form the outputs, with 1 corresponding to a connection and a 0 no-connection. It should be noted that the generators as published have their LSB (right-most bit) correspond to the most significant bit of the shift register. The MSB (left-most bit) gives the connection to the input.

Table 5.7 Formation of A Matrix

g_1	g_2	Modification to Element of A
False	False	No change
False	True	Multiply element by ω
True	False	Multiply element by ω
True	True	Multiply element by ω^2

Row 1 is unchanged. The other $2^{CL-2} - 2$ rows are changed as follows:

- An auxiliary table with $2 * (2^{CL-2} - 2)$ entries is formed.
- The next state (starting with state 2) is listed for two rows.
- Listed next on each row is the state from which the next state is reached.
- The next element on the row is the input that causes the current state to transition to the next state.
- The next two entries on each row are the outputs of the two generator functions, g_1 and g_2.

These bits are then generated as follows:

- The state and its input are considered as a CL component vector.
- Each element is logically ANDed with its corresponding position of the generator function to form a new CL length logical vector.
- These elements are exclusive ORed to determine g_1 and g_2.
- The matrix **A** is then modified according to Table 5.7.

Note that if the auxiliary table is formed first, the contribution to the elements of A by the output can be entered at the same time as the input contribution. Table F.1 in Appendix F shows a table of the A matrix for CL = 5.

The remaining steps in determining the performance of the code are mechanical:

1. Form $I - A$
2. Calculate $[I - A]^{-1}$
3. Form $T[\omega, I] = \mathbf{c} \cdot [\mathbf{I} - \mathbf{A}]^{-1}\mathbf{b}$
4. Calculate $\frac{\partial T[\omega, I]}{\partial I}$

5. Evaluate $\frac{\partial T[\omega, I]}{\partial I}$ at $I = 1$ and $\omega = \exp\left[-.5 * \frac{E_b}{N_o}\right]$

6. $P_b \leq \frac{1}{\left(\pi * d_{free} * \frac{E_b}{N_o}\right)} \frac{\partial T[\omega, I]}{\partial I}\Bigg|_{\substack{I=1 \\ \omega=\exp\left[-.5*\frac{E_b}{N_o}\right]}}$

Figure 5.24 shows plots of P_b for the uncoded case, and for CL = 3 and CL = 4. Note the importance of the coding (and higher coding gain) as performance (lower bit rate) is increased. Figure 5.25 shows the same plots for CL = 5, CL = 6, and CL = 7. Although each increase in the constraint length improves performance, beyond CL = 7 there is a diminishing return.

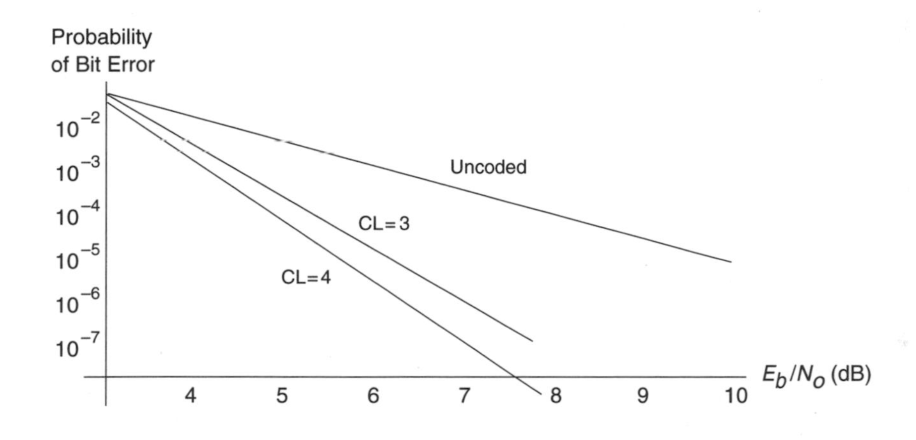

Figure 5.24 Probability of Bit Error Versus E_b/N_o for Rate 1/2 Code for Uncoded, CL = 3 and CL = 4

5.7.3 Punctured Codes

Any rational fraction code rate can be achieved with a shift register and appropriate exclusive OR generators: k bits are shifted into the shift register and η bits of output are generated from the exclusive OR circuits. Such codes were created this way originally.

However, this approach really requires a different encoder for each code, although both DSS and DVB support multiple code rates. Thus, it is important that a single encoder be employed. The solution is punctured codes.

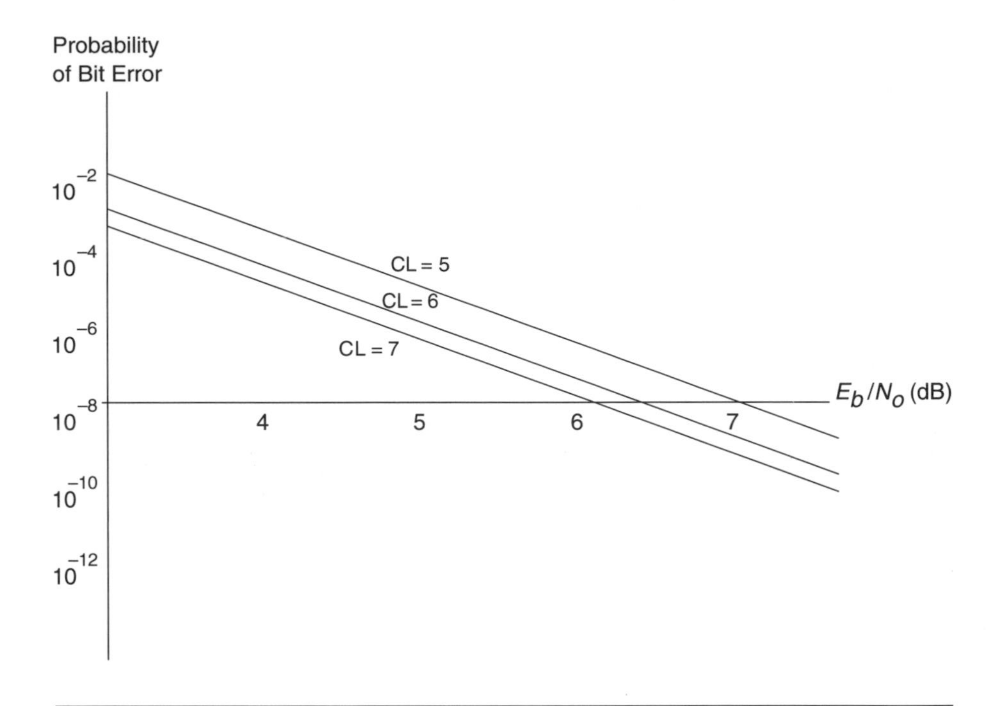

Figure 5.25 Probability of Bit Error, for CL = 5, 6, and 7

The type of punctured code important for DBS is to have a basic rate 1/2 encoder that generates two output bits for each input bit. Take the case for a rate 2/3 code: Two bits are shifted into the shift register and four output bits are generated. One of the output bits is discarded, leaving three bits to be transmitted. The discarded bit is the second bit from the first generator.

Table 5.8 shows the puncturing pattern for the convolutional codes of interest to DBS. A zero in Table 5.8 indicates the bit is not transmitted.

Table 5.8 Puncturing Patterns

	Code Rate				
	2/3	3/4	5/6	6/7	7/8
Generator 1	10	101	10101	100101	1000101
Generator 2	11	110	11010	111010	1111010

5.7.4 Performance of (3, 2, 7) Punctured Code and Concatenated Code

Finally, we can give the overall bit error rate performance of the concatenated code. It can be shown that

$$P_b < \frac{2^{k-1}}{2^k - 1} \sum_{j=t+1}^{n} \frac{j+t}{n} \binom{n}{j} P_s^j (1-P_s)^{n-j} \tag{5.37}$$

where k = bits per RS symbol
n = RS # symbols
t = RS error correcting capability
P_s = bit error rate into RS decoder from the inner coder

For both DSS and DVB, $k = 8$, and $t = 8$, so

$$P_b < \frac{2^7}{255} \sum_{j=9}^{n} \frac{j+8}{n} \binom{n}{j} P_s^j (1-P_s)^{n-j}$$

$$< .50196 \sum_{j=9}^{n} \frac{(j+8)}{n} \binom{n}{j} P_s^j (1-P_s)^{n-j}$$

$$(1-P_s)^{n-j} = (1-P_s)^n \cdot (1-P_s)^{-j}$$

$$= \left(\frac{(1-P_s)^n}{(1-P_s)^j} \right)$$

This leads to

$$\therefore P_b \le \frac{.50196}{n} * (1-P_s)^n * \left(\sum_{j=9}^{n} (j+8) * \binom{n}{j} * \left(\frac{P_s}{1-P_s} \right)^j \right) \tag{5.38}$$

For DSS, $n = 146$,

$$\therefore P_b \le \frac{.50196}{146} * (1-P_s)^{146} * \left(\sum_{j=9}^{146} (j+8) * \binom{146}{j} * \left(\frac{P_s}{1-P_s} \right)^j \right) \tag{5.39}$$

If the results from the (3, 2, 7) concatenated code are used to generate P_s in Equation (5.39), the curve shown in Figure 5.26 results.

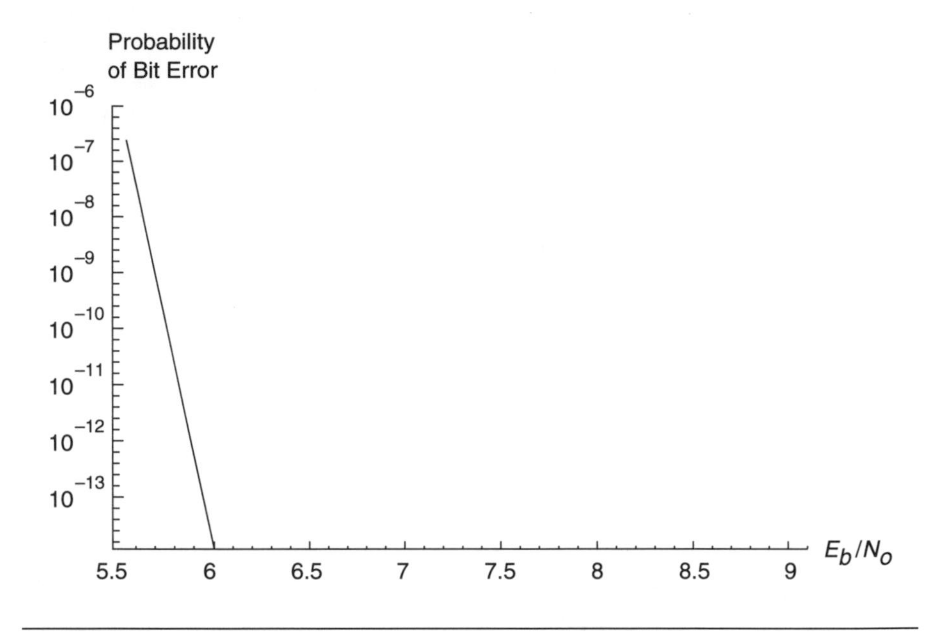

Figure 5.26 Probability of Bit Error for (3, 2, 7) Convolutional Code Concatenated with (146, 130) Reed-Solomon Code

The striking thing about Figure 5.26 is that the curve is essentially vertical: If E_b/N_o is less than 5.5 dB, the performance degrades to useless. On the other hand, if E_b/N_o is greater than 6 dB, there are essentially no bit errors.

References

[Biglieri+91] Biglieri, Ezio, Dariush Divsalar, Peter J. McLane, and Marvin K. Simon, *Introduction to Trellis Coded Modulation.* New York: Macmillan, 1991.

[Clark+81] Clark, George C. Jr., and J. Bibb Cain, *Error-Correction Coding for Digital Communication.* New York: Plenum Press, 1981.

[Couch97] Couch, Leon W. II, *Digital and Analog Communication Systems,* 4th ed. Upper Saddle River, NJ: Prentice Hall, 1997.

[Daut+82] Daut, David G., James W. Modestino, and Lee D. Wismer, New short constraint length convolutional code constructions for selected rational rates, *IEEE Transactions on Information Theory,* IT-28(5), 1982.

[Forney73] Forney, G. David, The Viterbi algorithm, *Proceedings of the IEEE,* 61(3), 1973.

[Lin+83] Lin, Shu, and Daniel J. Costello, Jr., *Error Control Coding: Fundamentals and Applications.* New York: Prentice-Hall Publishing, 1983.

[Proakis89] Proakis, John G., *Digital Communications,* 2nd ed. New York: McGraw-Hill, 1989.

[Ramsey70] Ramsey, John L., Realization of optimum interleaves, *IEEE Transactions on Information Theory,* IT-16(3), 1970.

[Reed+60] Reed,. I. S., and G. Solomon, Polynomial codes over certain finite fields, *Journal Society of Industrial Applied Mathematics,* 8(2), 1960.

[Viterbi95] Viterbi, Andrew J., *CDMA Principles of Spread Spectrum Communication.* Reading, MA: Addison-Wesley, 1995.

[Wilson96] Wilson, Stephen G. *Digital Modulation and Coding.* Upper Saddle River, NJ: Prentice Hall, 1996.

CHAPTER SIX

Conditional Access

The providers of program materials and the distributor of the programs (the DBS operator) understandably want to be paid for their intellectual property and services. To ensure that the broadcast of program materials can be viewed only by those who are authorized for such viewing (and who have agreed to pay for such services), conditional access (CA) systems have been developed. As the name implies, access to the materials is conditional, based on the viewers being authorized to view them.

6.1 Objectives of a CA System for DBS

A DBS CA system should be able to provide the following:

1. Access control
2. Copy control (antitaping)
3. Regional control (regional blackout)
4. User control (parental lockout)
5. Addressing capabilities to handle 100 million subscribers

Part of the CA system should permit a backlink from the subscriber to the management center so that a viewing history can be retrieved from the Integrated Receiver Decoder (IRD), thus permitting impulse pay-per-view. This link should provide a digital signature or authentication that the correct IRD is responding.

The CA system also must be capable of recovering from security compromise.

6.2 Types of Attackers

In the popular media, those who try to breach CA systems in an effort to view the materials for free are called pirates. Those who succeed are actually thieves, but in this book they will be called attackers.

There are two basic types of attackers: the graduate student who attacks the CA system because of the challenge and the commercial attacker who hopes to sell illegal access. The first, or "lone-wolf" variety, poses little threat because there are so few of them (fewer than 1 or 2 percent of those involved). The commercial attacker is another story.

Make no mistake—piracy is potentially a big business. To illustrate, suppose a DBS provider has 10 million users, but 10 percent of them, or one million, are provided access by an attacker. If the average subscriber pays $40 per month, or $480 per year, the loss to the DBS provider is $480 million per year! Even if the attacker charges only half that of the legitimate provider, his revenues for the year are still $240 million, almost all of which is pure profit. It is not surprising that serious attacks are made on the DBS CA system and that DBS operators make a major effort to defeat the attackers.

6.3 Some Encryption Preliminaries

Because all DBS systems involve the transmission of digital bit streams, the corresponding CA systems are best implemented by encryption/decryption technology (as opposed to scrambling, and so on). Thus, DBS systems can take advantage of the huge effort that has gone into the cryptography field.

The creation and breaking of codes is probably as old as human history. For a fascinating review of this history, see David Kahn's book, *The Codebreakers* [Kahn67].

All modern encryption algorithms employ a key, denoted by k. Using E to denote encryption, D to denote decryption, P to denote the plaintext message to be sent, and C for the coded message, we can write

$$C = E_k(P) \tag{6.1}$$

$$P = D_k(C) \tag{6.2}$$

In words, the coded or cipher message C is created by encrypting the plaintext message P with the key k. The plaintext message P is recovered from C by decrypting it with the same key k.

6.4 Mathematical Preliminaries

The following sections require an understanding of modulo arithmetic. The basic notation is

$$a = b \bmod n, \text{if } a = b + k * n \tag{6.3}$$

where k is an integer,

$$k = 0, 1, 2, 3, \ldots,$$

and $*$ denotes multiplication.

Another way of saying this is that a is the remainder after b is divided by n. It is easy to show the following modulo arithmetic operations:

$$\begin{aligned}(a + b) \bmod n &= (a \bmod n + b \bmod n) \bmod n \\ (a * b) \bmod n &= ((a \bmod n) * (b \bmod n)) \bmod n \\ (a * (b + c)) \bmod n &= ((a * b) \bmod n + (a * c) \bmod n) \bmod n\end{aligned} \tag{6.4}$$

Powers in modulo arithmetic are just extensions of multiplication.

The most important property of modulo arithmetic used in the following sections is that

$$[b \bmod n]^y \bmod n = b^y \bmod n \tag{6.5}$$

For example, let $b = 7, n = 5$, and $y = 4$. Working on the left side of (6.5),

$$[b \bmod n]^y \bmod n = [7 \bmod 5]^4 \bmod 5 = 2^4 \bmod 5 = 1$$

The right-hand side is

$$7^4 \bmod 5 = 2401 \bmod 5 = 1$$

In modulo arithmetic, division is defined as a multiplicative inverse. That is, if $(a * c) \bmod n = 1$, then c is the multiplicative inverse of a and is written $c = a^{-1}$. As opposed to finding multiplicative inverses with real or rational numbers, multiplicative inverses in modulo arithmetic may not exist.

Even when multiplicative inverses exist, they are not easy to calculate. It is interesting that an extension of an algorithm attributed to Euclid in 300 BC (although it may be several hundred years older) can be used to find modulo multiplicative inverses.

6.4.1 The Euclidean Algorithm

The Euclidean Algorithm finds the greatest common divisor of two numbers, say r_0 and r_1, noted gcd (r_0, r_1), where $r_0 > r_1$. The Euclidean Algorithm then forms

$$\begin{aligned} r_0 &= q_1 * r_1 + r_2 && 0 < r_2 < r_1 \\ r_1 &= q_2 * r_2 + r_3 && 0 < r_3 < r_2 \\ &\vdots \\ r_{m-2} &= q_{m-1} * r_{m-1} + r_m && 0 < r_m < r_{m-1} \\ r_{m-1} &= q_m * r_m \end{aligned}$$

Then, gcd $(r_0, r_1) = r_m$.

It can be shown that an element has a multiplicative inverse modulo r_0 if and only if gcd $(r_0, r_1) = 1$. Thus, in Euclid's Algorithm, r_1 has a multiplicative inverse if and only if $r_m = 1$.

Euclid's Algorithm shows whether an inverse exists but does not calculate the inverse. The following Extended Euclid's Algorithm [Stinson95] actually calculates the inverse.

6.4.2 Extended Euclid's Algorithm

Define a sequence of numbers $t_0, t_1, \ldots t_m$ according to the following recursion:

$$\begin{aligned} t_0 &= 0 \\ t &= 1 \\ &\vdots \\ t_j &= t_{j-1} - q_j t_{j-1} \bmod r_0, \text{ if } j = 2 \end{aligned}$$

where the q_j are defined above. Then it can be shown for $0 \leq j \leq m$, that $r_j = t_j r_1 \bmod r_0$. If gcd $(r_0, r_1) = 1$, then $t_m = r_1^{-1} \bmod r_0$. This can be put into the following algorithmic form:

1. $n_0 = n$
2. $b_0 = b$
3. $t_0 = 0$

4. $t = 1$

5. $q = \left\lfloor \frac{n_o}{b_0} \right\rfloor$

6. $r = n_0 - q * b_0$

7. While $r > 0$, do

8. temp $= t_0 - q * t$

9. If temp $= 0$, then temp $=$ temp mod n

10. If temp ≤ 0, then temp $= n - ((-\text{temp}) \bmod n)$

11. $t_0 = t$

12. $t =$ temp

13. $n_0 = b_0$

14. $b_0 = r$

15. $q = \left\lfloor \frac{n_o}{b_0} \right\rfloor$

16. $r = n_0 - q * b_0$

17. If $b_0 \neq p1$, then b has no inverse mod n.

Else, $b^{-1} = t \bmod n$.

Example Calculate $28^{-1} \bmod 75$

$$n_0 = 75$$
$$b_0 = 28$$
$$t_0 = 0$$
$$t = 1$$
$$q = \left\lfloor \frac{n_o}{b_0} \right\rfloor = \left\lfloor \frac{75}{28} \right\rfloor = 2$$
$$r = 75 - 28 * 2 = 19$$

Start recursion:

1. temp $= 0 - 2 * 1 = -2$
   temp $= n - ((-(-2)) \bmod 75) = 75 - 2 = 73$

$t_0 = 1$

$t = 73$

$n_0 = 28$

$b_0 = 19$

$$q = \left\lfloor \frac{28}{19} \right\rfloor = 1$$

$r = 28 - 1 * 19$

2. temp $= 1 - 1 * 73 = 72$

   temp $= 75 - ((-(-72)) \bmod 75) = 3$

   $t_0 = 73$

   $t = 3$

   $n_0 = 19$

   $b_0 = 9$

$$q = \left\lfloor \frac{19}{9} \right\rfloor = 2$$

$r = 19 - 2 * 9 = 1$

3. temp $= 73 - 2 * 3 = 67$

   $t_0 = 3$

   $t = 67$

   $n_0 = 9$

   $b_0 = 1$

$$q = \left\lfloor \frac{9}{1} \right\rfloor = 9$$

$r = 9 - 9 * 1 = 0$

Since $r = 0$, the recursion terminates. Since $b_0 = 1$, an inverse exists and t is that inverse. Thus, $28^{-1} \bmod 75 = 67$. For the doubters, calculate $26 * 67 = 1876$, $1876/75 = 25$ with $r = 1$.

6.5 Cryptographic Algorithms

A DBS/CA system will typically use four or five different cryptographic techniques to perform the various required functions. A symmetric key technique, such as the Data Encryption Standard (DES) discussed in the following section, can provide high-speed encryption that can be used to encrypt the individual services. However, the key distribution can be a problem. A public key system such as RSA solves the key-distribution problem but is too slow to encrypt services. Digital Signature algorithms can prove that the source of a communication is valid, and message authentication codes (MAC) can be used to ensure that a message has not been tampered with.

In this section the key algorithms are described. Section 6.7 presents a generic CA system that employs these algorithms.

6.5.1 The Data Encryption Standard

DES is a U.S. national encryption standard based on original input from IBM. The National Security Agency (NSA) consulted with the National Bureau of Standards (NBS), now the National Institute of Standards and Technology (NIST), on the development of the standard.

In considering cryptographic algorithms for DBS/CA, there are three criteria:

1. The algorithm must not cause an expansion of the data.
2. The algorithm must be capable of very fast operation.
3. The decryption must be capable of implementation in very inexpensive hardware.

The DES algorithm satisfies all of these criteria.

DES is a block cipher. It takes in 64 bits, encrypts the 64 bits with a 56-bit key, and outputs 64 bits of cipher text. Thus, we see that the first criterion is met; the cipher text length is exactly the same as the plaintext length (zero overhead).

Note: The ciphertext length can never be less than the plaintext because the plaintext could never be recreated. It is a basic theorem of linear algebra that an n-to-less-than-n mapping cannot be inverted.

The DES algorithm puts the 64 bits of input through 16 identical iterating rounds of processing. The processing is shown in Figures 6.1 and 6.2.[1]

---

[1] Figures 6.1 and 6.2 and Tables 6.1 through 6.5 were adapted from B. Schneier, *Applied Cryptography* [Schneier94], as has been documented in the applicable U.S. standard.

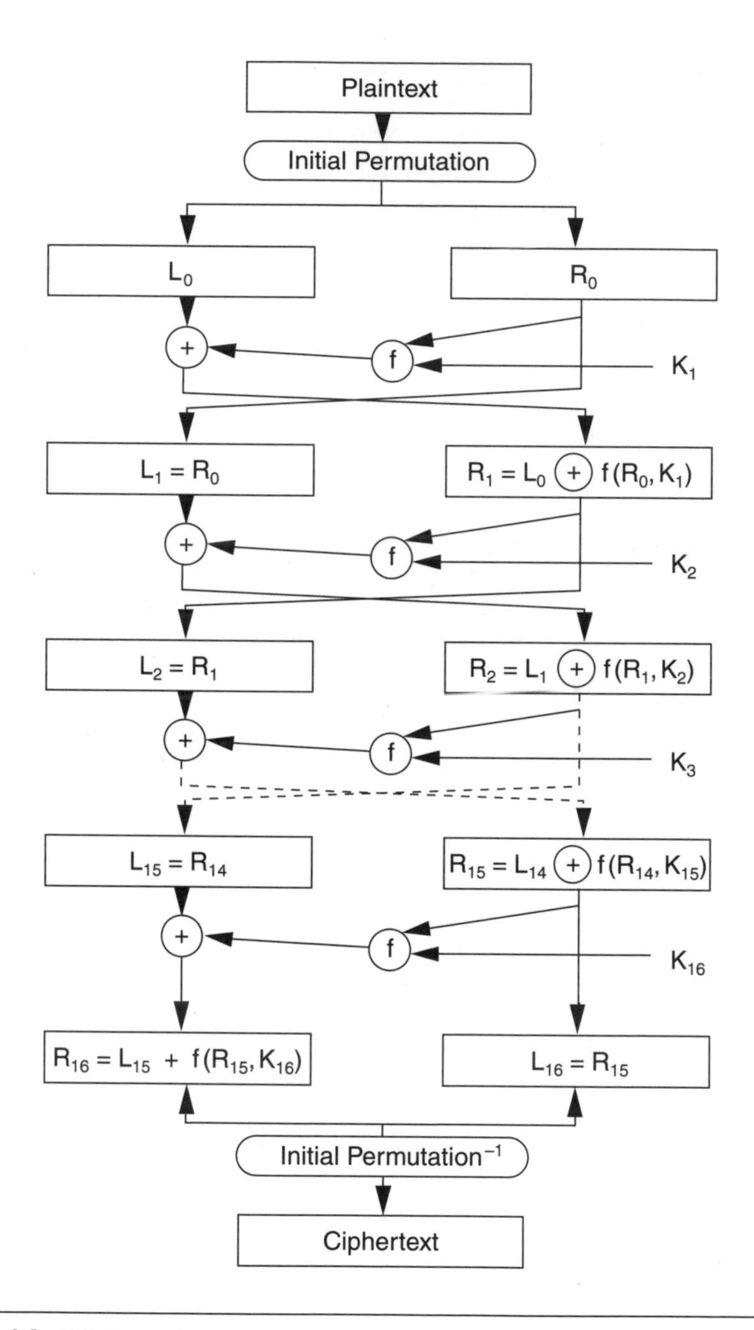

Figure 6.1 DES Processing

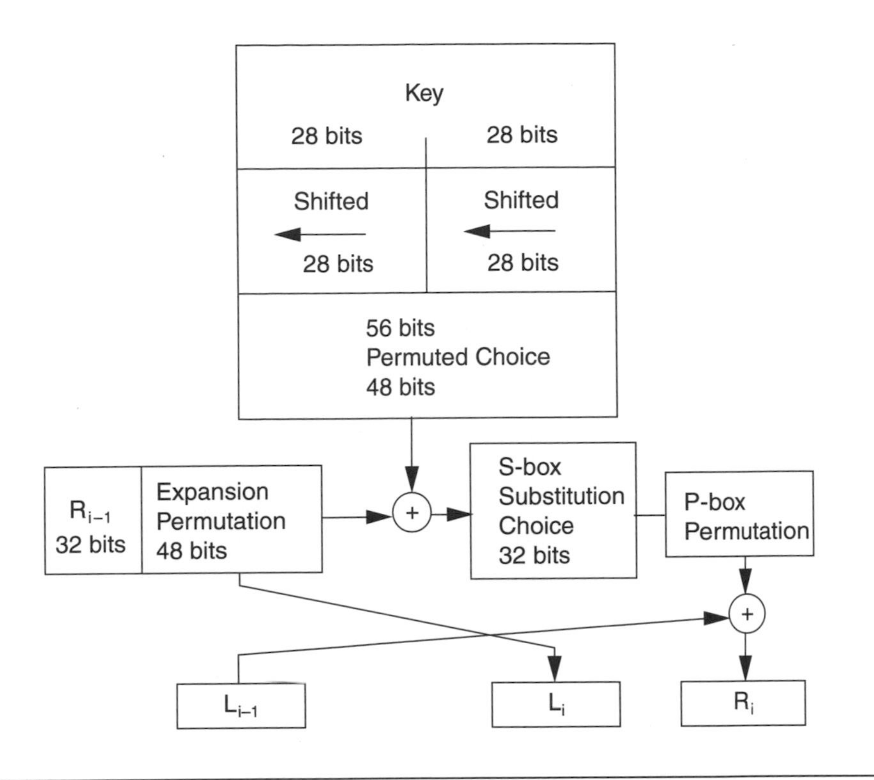

Figure 6.2 One Round of DES

Referring to Figure 6.2, the 56-bit key is divided into two 28-bit halves. The halves are shifted left one or two bits, depending on the round. Table 6.1 shows the number of shifts per round. After being shifted, 48 of the 56 bits are selected according to Table 6.2. This is called the compression permutation.

Referring to Figure 6.2, the 32 bits of the right half of the data are expanded to 48 bits in an expansion permutation. The mapping is shown in Table 6.3.

Table 6.1 Key Shifts per Round

Round	1	2	3	4	5	6	7	8	9	10	11	12	13	14	15	16
Number of Shifts	1	1	2	2	2	2	2	2	1	2	2	2	2	2	2	1

Table 6.2 Compression Permutation

14	17	11	24	1	5	3	28	15	6	21	10
23	19	12	4	26	8	16	7	27	20	13	2
41	52	31	37	47	55	30	40	51	45	33	48
44	449	39	56	34	53	46	42	50	36	29	32

Table 6.3 Expansion Permutation

32	1	2	3	4	5	4	5	6	7	8	9
8	9	10	11	12	13	12	13	14	15	16	17
16	17	18	19	20	21	20	21	22	23	24	25
24	25	26	27	28	29	28	29	30	31	32	1

Much of DES's design revolves around reaching—as quickly as possible—the condition of having every bit of the cipher text depend on every bit of the plaintext and every bit of the key. The expansion permutation contributes to this.

After the compressed key is exclusive ORed with the expanded block, the 48-bit result moves to a substitution operation. The substitution is performed by eight substitution boxes or *S*-boxes.

The 48 bits are divided into eight blocks of six bits each. Each of the blocks is associated with one of the *S*-box tables. Each table consists of four rows and 16 columns. Table 6.4 shows all eight boxes.

If the six bits in a block are labeled b_1, b_2, b_3, b_4, b_5, and b_6, then b_1 and b_6 form a two-bit row address. The four bits b_2, b_3, b_4, and b_5 form a four-bit column address. The output is the corresponding four-bit entry from the *S*-box. The eight four-bit entries from the *S*-box outputs are concatenated to form a new 32-bit output.

The 32-bit output of the *S*-boxes becomes the input to the permutation or *P*-box. Table 6.5 shows the one-for-one permutations of the *P*-box. Before the first round and after the sixteenth round, there is an initial permutation and a final permutation. Since these do not add to the cryptographic security of the algorithm, they are not discussed here.

One of the remarkable aspects of the DES algorithm is that the same algorithm works for decrypting and encrypting if the order of the keys used for each round is reversed.

Table 6.4 *S*-Boxes

S-Box 1															
14	4	13	1	2	15	11	8	3	10	6	12	5	9	0	7
0	15	7	4	14	2	13	1	10	6	12	11	9	5	3	8
4	1	14	8	13	6	2	11	15	12	9	7	3	10	5	0
15	12	8	2	4	9	1	7	5	11	3	14	10	0	6	13
S-Box 2															
15	1	8	14	6	11	3	4	9	7	2	13	12	0	5	10
3	13	4	7	15	2	8	14	12	0	1	10	6	9	11	5
0	14	7	11	10	4	13	1	5	8	12	6	9	3	2	15
13	8	10	1	3	15	4	2	11	6	7	12	0	5	14	9
S-Box 3															
10	0	9	14	6	3	15	5	1	13	12	7	11	4	2	8
13	7	0	9	3	4	6	10	2	8	5	14	12	11	15	1
13	6	4	9	8	15	3	0	11	1	2	12	5	10	14	7
1	0	13	0	6	9	8	7	4	15	14	3	11	5	2	12
S-Box 4															
7	13	14	3	0	6	9	10	1	2	8	5	11	12	4	15
13	8	11	5	6	15	0	3	4	7	2	12	1	10	14	9
10	6	9	0	12	11	7	13	15	1	3	14	5	2	8	4
3	15	0	6	10	1	13	8	9	4	5	11	12	7	2	14
S-Box 5															
2	12	4	1	7	10	11	6	8	5	3	15	13	0	14	9
14	11	2	12	4	7	13	1	5	0	15	10	3	9	8	6
4	2	1	11	10	13	7	8	15	9	12	5	6	3	0	14
11	8	12	7	1	14	2	13	6	15	0	9	10	4	5	3
S-Box 6															
12	1	10	15	9	2	6	8	0	13	3	4	14	7	5	11
10	15	4	2	7	12	9	5	6	1	13	14	0	11	3	8
9	14	15	5	2	8	12	3	7	0	4	10	1	13	11	6
4	3	2	12	9	5	15	10	11	14	1	7	6	0	8	13

(continued)

Table 6.4 Continued

S-Box 7															
4	11	2	14	15	0	8	13	3	12	9	7	5	10	6	1
13	0	11	7	4	9	1	10	14	3	5	12	2	15	8	6
1	4	11	13	12	3	7	14	10	15	6	8	0	5	9	2
6	11	13	8	1	4	10	7	9	5	0	15	14	2	3	12
S-Box 8															
13	2	8	4	6	15	11	1	10	9	3	14	5	0	12	7
1	5	13	8	10	3	7	4	12	5	6	11	0	14	9	2
7	11	4	1	9	12	14	2	0	6	10	13	15	3	5	8
2	1	14	7	4	10	8	13	15	12	9	0	3	5	6	11

Table 6.5 *P*-Box Permutations

16	7	20	21	29	12	28	17	1	15	23	26	5	18	31	10
2	8	24	14	32	27	3	9	19	13	30	6	22	11	4	25

6.5.2 Public Key Cryptography

For an economically viable DBS service, a number of different services must be offered. Each service costs extra and requires a different encryption key. Further, it is desirable to frequently change the keys for each service.

Public Key Cryptography provides a means for delivering these keys to the subscriber. In public key cryptograph there are two keys: the public key and the private key. Neither can be calculated from the other.

Suppose the conditional access center has a database of the public key associated with each subscriber's IRD. Suppose further that there are two very large integers, n and g, with $n > g$. Both n and g can be publicly known without compromising security. Suppose x is the subscriber's public key and y is the subscriber's private key, which is embedded in the IRD.

The CA center forms $X = g^x \bmod n$, and transmits it through the satellite to the subscriber's IRD. The IRD calculates $k = X^y \bmod n$, or

$$k = [g^x \bmod n]^y \bmod n \tag{6.6}$$

Using Equation (6.5), this reduces to

$$k = g^{x*y} \bmod n \tag{6.7}$$

Thus, even though an attacker knows g, n, and X, he still cannot determine k.

A practical public key algorithm for the distribution of DES keys is the RSA algorithm, named for its three inventors: Ronald Rivest, Adi Shamir, and Leonard Adelman.

Choose two large prime numbers, p and q (which must remain secret), and compute

$$n = p * q \tag{6.8}$$

Then randomly choose the public key e such that e and $(p - 1) * (q - 1)$ are relatively prime. Compute the private key $d = e^{-1} \bmod ((p - 1) * (q - 1))$.

The encryption formulas are then

$$c = M^e (\bmod n) \tag{6.9}$$

which is transmitted. To decode,

$$M = c^d (\bmod n) \tag{6.10}$$

This can be derived by the following:

$$\begin{aligned} c^d (\bmod n) &= (M^e)^d (\bmod n) \\ &= M^{e*d} (\bmod n) \end{aligned} \tag{6.11}$$

from Equation (6.5).

Now, $e * d = k * (p - 1) * (q - 1) + 1$ from the definition of modulo arithmetic and Equation (6.5).

$$\begin{aligned} c^d (\bmod n) &= M^{k*(p-1)*(q-1)+1} \\ &= M * M^{k*(p-1)*(q-1)} \end{aligned} \tag{6.12}$$

However, since $M^{k(p-1)(q-1)} = 1$, $c^d (\bmod n) = M$.

Note that we can continue to send messages to the IRD as fast as the RSA algorithm can decode them. These messages may be the DES keys or other instructions, such as "call in on the modem."

6.5.3 One-Way Hash Functions

In many cases, particularly when dealing with long messages, it is important to determine whether the messages are authentic without dealing with the messages themselves. These functions generally are compactions of the messages. They have many names, including message digest and message authentication.

A one-way hash function $H(M)$ operates on an arbitrary-length message M and produces a fixed-length hash value h of fixed-length n. One-way hash functions have the following properties:

- Given M, it is easy to compute h.
- Given h, it is hard to compute M.

Note that because h is a many-to-one mapping, in general more than one M can create h. Given M, it is hard to find another message M' such that $H(M') = H(M)$, even though we know they exist.

Most one-way hash functions are designed to take in two n-length inputs and produce a single n-length output. Usually, the inputs are a block of text, and the hash of the previous block of text, that is,

$$h_i = f(M_i, h_{i-1})$$

The hash of the last block is the hash for the entire message.

Professor Ronald Rivest (of RSA fame) has developed one-way hash functions called MD4 and MD5 (as in Message Digest), both of which produce 128-bit hash outputs. MD5 is proposed in the generic CA system discussed in section 6.7. However, the Secure Hash Standard, which is the United States national standard, has 160-bit hash outputs and, thus, is cryptographically stronger at the cost of more processing.

The reader interested in more information on one-way hash functions is referred to Schneier's excellent book [Schneier94], which is not only readable, but has an excellent bibliography of original source materials.

6.5.4 Digital Signatures

In certain electronic communication situations, it is desirable to have a digital signature, which plays the same role in electronic communication that a written signature plays in written communication. The digital signature of a message is a bit string attached to the message. Digital signatures are almost always implemented with public-key cryptographic techniques.

It is frequently more practical to sign a one-way hash of the message rather than signing the message itself.

Uriel Feige, Amos Fiat, and Adi Shamir (the S of RSA fame) developed an algorithm called a zero-knowledge proof of identity, which is an authentication/digital signature scheme. The basic concept of this scheme is covered by U.S. Patent 4,748,668. It is used in DBS/CA to prove that an inserted smart card is valid.

6.6 Generic CA System

Because information is a valuable tool for information thieves, we must not divulge the details of any specific CA system. What follows is a description of a generic CA system that is a variant of one that was proposed to the Digital Audio Video Council (DAVIC) in May of 1995 by Scientific Atlanta [DAVIC95].

Figure 6.3 shows a simplified block diagram of this generic CA system. At the transmitter, a random DES key is selected. Note that this is not a

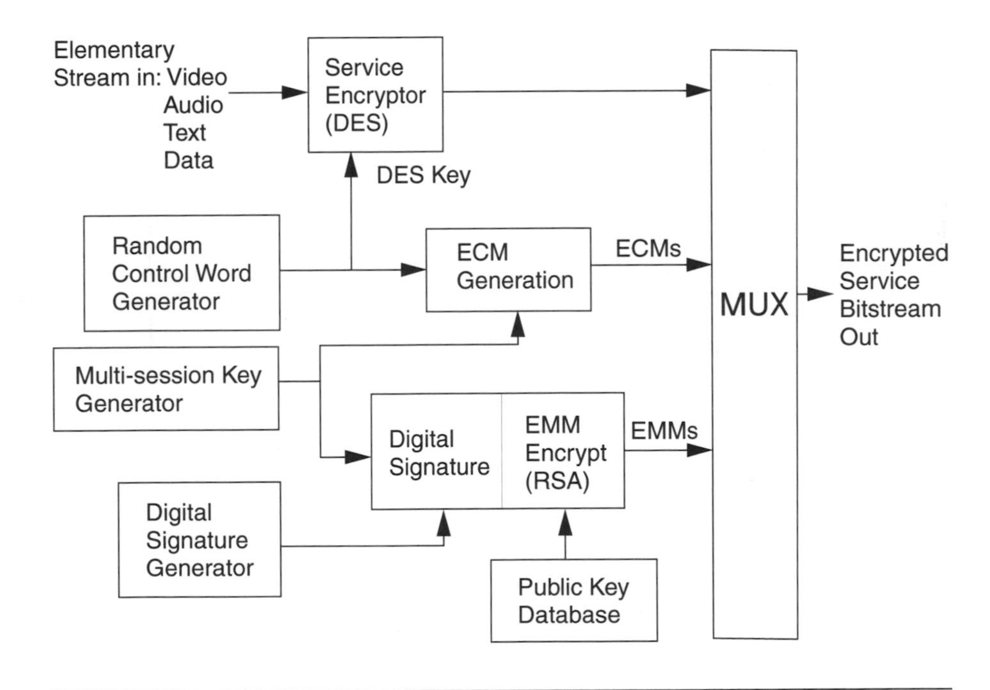

Figure 6.3 Generic CA Encoder

pseudorandom number but, rather, a truly random number that should be derived from a physical process, such as the noise in a diode, and so on. It is used to encrypt the video and audio for a service. The DES key becomes the message for encryption by the multisession key as part of the ECM generation process. The multisession key is signed with a digital signature and then encoded with the public key of an RSA public encryption system. The ECMs and the EMMs are multiplexed together to form the encrypted service bitstream.

Figure 6.4 shows the block diagram for the ECM generation. The DES key is encrypted with the multisession key using triple DES encryption. The multisession key, the DES key, and up to 320 bits of stream description information are combined as the "message" and authenticated by the MD5 message authentication code (MAC). The encrypted DES key, the MAC, and the stream description bits become the ECM.

Figure 6.5 shows triple DES encryption. The plaintext is first DES encrypted with key K1. It is then inverse DES encrypted with key K2. Finally, it is DES encrypted with key K1. In the decoder, the process is reversed. The motivation for triple DES is that it effectively increases the key length from 56 bits to 112 bits.

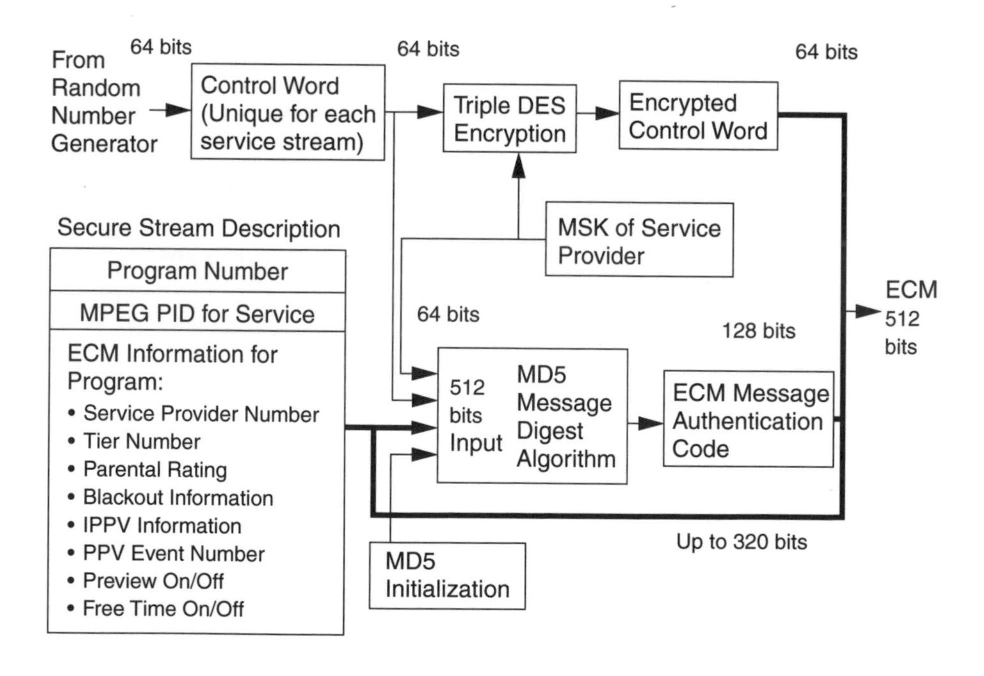

Figure 6.4 ECM Generator

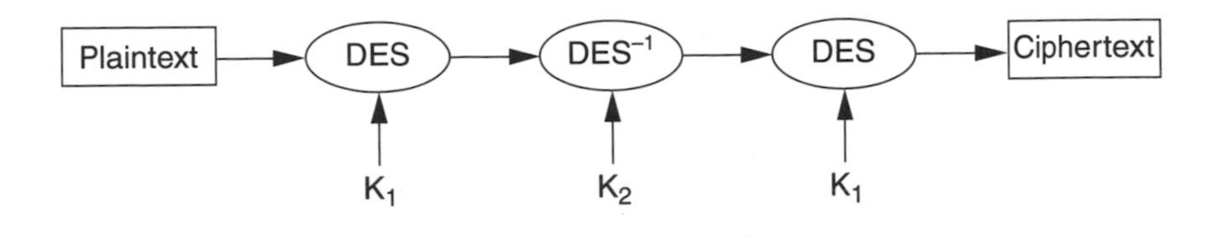

Figure 6.5 Encoding with Triple DES

CA Decoding

1. The encrypted service bitstream is demultiplexed and the EMMs are selected.
2. The EMM is decrypted with the private part of the RSA key.
3. Check the digital signature of the multisession key. If valid, proceed to 4.
4. Select ECM Check ECM MAC. If valid, stream description information is stored in appropriate places in IRD.
5. Use the multisession key to triple DES decrypt the DES key.
6. Use the DES key to decrypt the individual services.

References

[DAVIC95] Paper presented by Scientific Atlanta at the Digital Audio Video Council meeting, Rome, May 1995.

[Kahn67] Kahn, David, *The Code Breakers.* New York: Macmillan, 1967.

[Schneier94] Schneier, Bruce, *Applied Cryptography.* New York: John Wiley & Sons, 1994.

[Stinson95] Stinson, Douglas R., *Cryptography Theory and Practice.* Boca Raton, FL: CRC Press, 1995.

PART THREE

MPEG International Standards

Foreword

The second work item of the Moving Picture Experts Group (MPEG), popularly known as MPEG 2, was developed to deal with full-resolution pictures and multichannel audio. The Direct Broadcast Satellite (DBS) was certainly one of the major target applications.

In July 1990 at the MPEG meeting in Porto, Portugal, we began the effort to develop the MPEG 2 standard. It was at this meeting that I first met Don Mead, who was representing Hughes Electronics. Hughes had begun a program called Sky Cable (later named DIRECTV) and had issued a request for information (RFI) to the industry to provide information about the DBS requirements. Because of the technical relevance of the RFI it was agreed that this document should be entered as an official MPEG document for the Porto meeting so that the MPEG members could take into account that throughout the three years of very intense effort to develop the MPEG 2 standard, Dr. Mead was intimately involved. He made particular contributions to the program management aspects. He was in perfect agreement with me about the need to ensure that the standard stayed on track for meeting its scheduled obligations, particularly on the requirements and systems subgroups.

The three chapters of Part Three of this book show the particularization of MPEG 2 to the DBS requirement. In them you will find all the information required to understand the application of MPEG to DBS.

Clearly, MPEG 2 and DBS were synergistic: DBS created a market for digital television (i.e., for MPEG 2), while MPEG 2 was the enabling technology for DBS.

I am confident that you will find this part of the book both stimulating and instructive.

Leonardo Chiariglione
MPEG Convenor, ISO/IEC
JTC1/SC29/WG11

The Need for Standards

Most people are not aware of the effect technical standards have on their everyday lives. Consumers in the United States can go to a store and buy a television set from any number of manufacturers and be assured that when they take it home, both the set and the terrestrial broadcast or cable signals will be received and understood by the set they bought. This is because long ago an organization called the National Television Systems Committee established the standard that goes by its name—NTSC. Similarly, fax transmission was around for many years but was prohibitively expensive. In 1979, I remember wanting to send an electronic schematic from New York to a field engineer in Ireland, and Western Union was more than happy to do so for $2,500. It was cheaper to take the schematic in hand, fly to Ireland, and hand it to the field engineer than to send it by fax. It was not until the Group 3 facsimile standard was established and faxes could be received on any Group-3-compatible machine that faxing really took off. Quick standardization also permitted the rapid growth of the compact disc.

Standards Bodies

Standards are established by a very large number of bodies. However, three of them are important for the purposes of this book.

1. The International Telecommunications Union (ITU) is a United Nations chartered body. The ITU has two main arms: the ITU-R, which is the standards-setting body for radio, and the ITU-T, which is the standards-setting body for telecommunications. The members of the ITU are the national ministries of telecommunication and posts. Since the United States does not have a ministry of telecommunication and posts, the State Department serves as the U.S. representative.

2. The second major international standards-setting body is the International Standards Organization (ISO), which has its headquarters in Geneva, Switzerland. While the ITU has specialized in telecommunications standards, the ISO has tended to specialize in information technology (although ISO 9000, the standard for quality control, is probably its most famous standard).

3. The American National Standards Institute (ANSI) represents the United States in the ISO. Also, the State Department usually delegates ANSI as its representative for ITU functions. ANSI has the responsibility for developing U.S. national positions on standards activities and, in the case of ISO, of providing the delegate rosters.

International Standards Organization

Figure P3.1 shows the organization of ISO with regard to the Moving Picture Experts Group (MPEG). In conjunction with the International Electrotechnology Commission (IEC), ISO has formed the Joint Technical Committee 1 (JTC 1) for information systems. This is the top organization within the ISO standards-making hierarchy.

Table P3.1 shows the membership of JTC 1. Note that the 26 *P* members are official voting members, whereas the 10 *O* members are observers. Reporting to JTC 1 is Subcommittee 29 (SC 29), which is responsible for picture coding efforts. In the early days of MPEG, what is now SC 29 was Subcommittee 2. Finally, the Moving Picture Experts Group is Working Group 11 of SC 29. It should be noted that the Working Group 11 took its number when it was part of SC 2. Thus, not all of the numbers between 1 and 12 exist.

MPEG History

In the mid-to-late 1980s it became apparent that video was going to go digital. It also was clear that if international standards were not developed for digital video, there would be a "Tower of Babel" as proprietary standards proliferated; therefore, in 1988 ISO chartered MPEG, Working

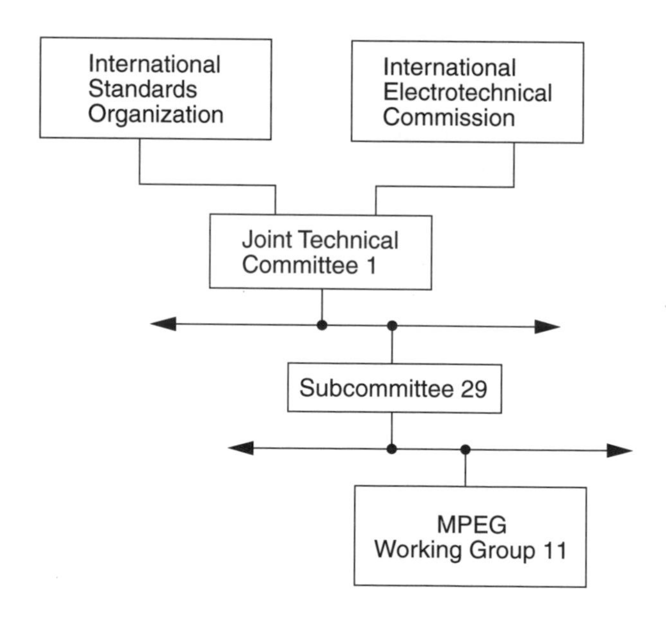

Figure P3.1 The Moving Picture Experts Group in the ISO

Table P3.1 Permanent and Observer Members of JTC 1

Signatory Nations		
Permanent Members		
Australia	Ireland	Russian Federation
Belgium	Israel	Singapore
Brazil	Italy	Spain
Czech Republic	Japan	Sweden
Denmark	Korea, Republic of	Switzerland
Finland	Netherlands	United Kingdom
France	Norway	United States
Germany	Portugal	
India	Romania	
Observer Members		
Austria	Hungary	Turkey
China	Indonesia	Yugoslavia
Greece	Poland	
Hong Kong	Slovakia	

Group 11, to develop standards for digital video. The following three work items were defined:

1. Video and associated audio at rates up to about 1.5 Mbps (later called MPEG 1)
2. Moving pictures and associated audio at rates up to about 10 Mbps (later called MPEG 2)
3. Moving pictures and associated audio at rates up to about 60 Mbps (later reduced to 40 Mbps and then cancelled)

MPEG 1 was oriented toward digital video stored on digital storage media (DSM). MPEG 2 was oriented toward broadcast media. MPEG 3 was for high-definition television (HDTV). As work progressed on the development of the standards, especially about the time that MPEG 1 was being completed and MPEG 2 was really in the mainstream, it became clear that the techniques employed in the standards could be used at any bitrate. While it was optimized around 1.5 Mbps, MPEG 1 could be used at very high bitrates. MPEG 2 could be used at virtually any bitrate.

Then a significant debate arose within the MPEG community as to whether to change the title of MPEG 2 by eliminating the 10 Mbps statement in its title. The U.S. delegation was able to convince its MPEG colleagues to do this, and the 10 Mbps was dropped from the MPEG 2 title. At the same time it was clear that the MPEG 2 could satisfy the needs of what originally had been thought would be a separate standard required for HDTV. Therefore, MPEG 3 was dropped and will never exist.

When the consortia of companies in the United States called the Grand Alliance selected a Video Compression standard for the U.S. HDTV service, they picked MPEG 2 video. This justifies the decision that MPEG 2 could provide the compression for HDTV.

MPEG Background

In 1988 the MPEG, Working Group 11, was formed under the convenorship of Dr. Leonardo Chiariglione of CSELT in Italy. The efforts of this group culminated in November 1992 when ISO Standard 11172, Parts 1, 2, and 3 were approved. The parts of the standard are as follows:

Part 1—Systems. This is the transport and multiplex layer, which provides for synchronization of the audio and video bitstreams.

Part 2—Video compression

Part 3—Audio compression

It should be noted that the three parts of the ISO standard are published in a stand-alone fashion so that each can be used by itself as well as with the other two parts.

The MPEG 2 effort was kicked off in July 1990 at the MPEG meeting in Portugal. It became ISO Standard 13818, Parts 1, 2, and 3, in November 1994 at the meeting in Singapore. The first three parts of MPEG 2 cover the same subjects as MPEG 1.

The first two chapters of Part Three of this book are based on MPEG 2. For reasons explained in Chapter 9, the Audio Compression Standard described is from MPEG 1. Copies of ISO Standards 11172 and 13818 can be obtained by contacting either of the following:

International Standards Organization, 1 Rue de Varembe, Case Postale 56, CH-1211, Geneva 20, SWITZERLAND (Tel: 22-749-0111).

Document Engineering Co., Inc. (DECO), 15210 Stagg Street, Van Nuys, CA 91405-1092 U.S.A.

CHAPTER SEVEN

MPEG 2 Systems

Part 1 of both ISO 11172 (MPEG 1) and ISO 13818 (MPEG 2) is called "Systems." It is easy to see the need for Video Compression (Part 2 in each standard) and Audio Compression (Part 3 in each standard); however, what is *Systems,* and why is it needed?

7.1 The Role of MPEG Systems

The video, audio, and other data for an MPEG-coded service must be multiplexed into a single bitstream. This is the first task of MPEG 2 Systems.

Next, the question is, when the multiplexer is receiving separately compressed video and audio bitstreams, how are they to be multiplexed so that the decoder can present them in a synchronized manner (for example, lip synchronization between audio and video)? A second role of Systems is to provide the means for such synchronization.

Even though MPEG bitstreams represent a continuous stream of bits, the bits need to be organized into groups (packets) so that bit errors cannot propagate beyond the borders of a single packet. In general, the longer the packet length, the more susceptible it is to bit errors. On the other hand, grouping the bits into packets creates overhead to accommodate the packet headers. In general, the shorter the packet length, the higher the overhead. There is, thus, a tradeoff in selecting a packet length between error resiliency and efficiency. In any event, forming the packets is a third function of MPEG Systems.

In most cases, decoders require Program Specific Information (PSI) to decode the arriving bitstream. Providing this PSI is MPEG Systems' fourth role. In summary then, MPEG Systems:

1. Multiplexes individual bitstreams into a single bitstream
2. Provides the means to synchronize the component bitstreams that create an audio and/or video service

3. Packetizes the bits into groups
4. Provides PSI

In MPEG 2 Systems, a program is defined as a set of meaningful Elementary Streams, such as audio and video, that have the same time base.

7.2 Transport Stream

Transport Streams are composed of Transport Stream packets with headers containing information that specifies the time when each byte is intended to enter a Transport Stream Decoder from the channel.

Through an additional multiplex-wide operation, a decoder is able to establish what resources are required to decode a Transport Stream. The Transport Stream contains information that identifies the pertinent characteristics and relationships between the Elementary Streams that constitute each program. Such information may include the language spoken in audio channels, as well as the relationship between video streams when multilayer video coding is implemented.

Figure 7.1 illustrates how a Transport Stream is constructed in two layers: a system layer and an elementary stream (ES) layer. The Transport Stream encoder wraps a system layer about an ES layer.

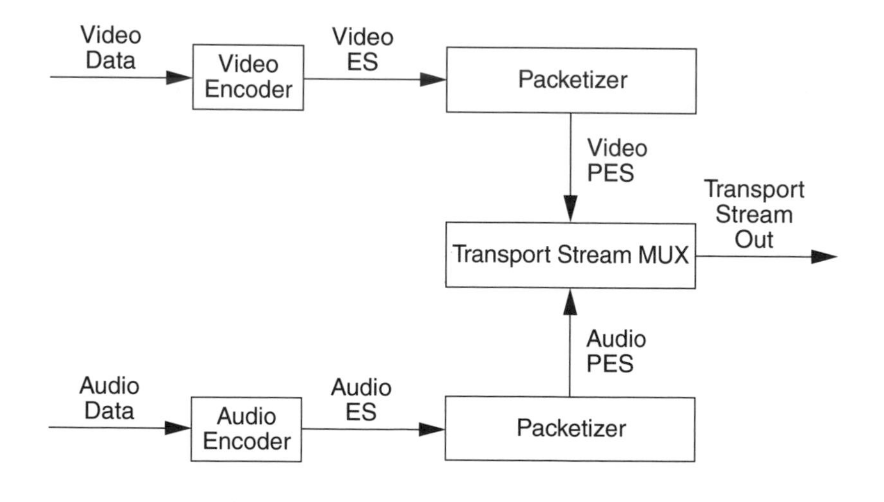

Figure 7.1 MPEG 2 Systems Transport Stream Encoding

The structure of a Transport Stream packet creates a number of design decisions. On one hand, the length of the packer header wants to be as small as possible in order to minimize the header overhead. On the other hand, the control features want to be as rich as possible, which implies a long header.

MPEG 2 Systems is able to achieve both of these seemingly contradictory objectives in the following manner:

- For the delivery of payload data, a fixed four-byte header is used. However, 2 bits (out of the 32) identify whether the information in the packet is payload data, or an Adaptation field.
- If an Adaptation field is indicated, a number of special parameters can be included in the packet payload. Since Adaptation fields are sent infrequently, the effect on the overall header efficiency is minimal. The Adaptation field itself is hierarchical in that it contains several additional sublayers of Adaptation fields.

Transport Streams are an envelope for constituent ESs. However, the Elementary Streams are usually not independent of each other. For example, an audio and/or video service will have an audio ES and a video ES. These two Elementary Streams are associated and are called a *program,* not unlike the language we would use for a "TV program." The constituent ESs of each program must have the same time base.

Elementary Stream data is carried in Packetized Elementary Stream (PES) packets. A PES packet consists of a PES packet header followed by data. Packetized Elementary Stream packets are inserted into Transport Stream packets. The first byte of each PES packet header is located at the first available payload location of a Transport Stream packet. A discussion of the PES packet structure is contained in a separate section that follows.

The Program Specific Information tables are carried in the Transport Stream. There are four PSI tables:

- Program Association Table
- Program Map Table
- Conditional Access Table
- Network Information Table

These tables contain information needed to demultiplex and present programs. The Program Map Table specifies other information about which packet IDs (PIDs), and, therefore, which Elementary Streams are associated to form each program.

7.2.1 Packet Length

Because its main use was intended to be with digital storage media (DSM) with attendant very low error rates, MPEG 1 Systems permitted very large (and variable) packet lengths. MPEG 2 Program Streams are designed for the same type of applications. On the other hand, one of the main applications for MPEG 2 was broadcast and other error-prone transport. Thus, the Transport Stream part of MPEG 2 Systems is appropriate for DBS and is covered in the remainder of this chapter.

The Transport Stream is comprised of a sequence of fixed-length packets of 188 bytes each. Note that 188 bytes is not an obvious length unless several special factors are considered. In a shortened Reed-Solomon block error-correction code, 188 is a natural size for the number of information bytes. To see this, recall that a t error-correcting Reed-Solomon code is defined by

Block length: $n = q - 1$

Number of parity check digits: $n - k = 2 * t$

Minimum distance: $d_{\min} = 2 * t + 1$

If we let $q = 2^8 = 256$, then $n = 255$. If we further let $t = 8$ so that eight bytes worth of errors can be corrected, a (255, 239) code results. This has been a U.S. National Aeronautics and Space Administration (NASA) standard Reed-Solomon code for a number of years. We can form a code with $n' = n - l$ and $k' = k - l$, which is known as a shortened Reed-Solomon code. By letting $l = 51$, a shortened Reed-Solomon code (204, 188) results.

A (204, 188) Reed-Solomon code is used for the Digital Video Broadcast (DVB) standard, for example. It will correct eight byte errors in each 204-byte packet. Thus, there is no overhead (other than the parity bytes) from the 188 information bytes to the 204 total bytes after application of the outer code.

The second reason for the 188 arises from the length of Asynchronous Transfer Mode (ATM) packets. Figure 7.2 shows the ATM packet structure. The structure consists of a 5-byte header, a 1-byte ATM Adaptation Layer

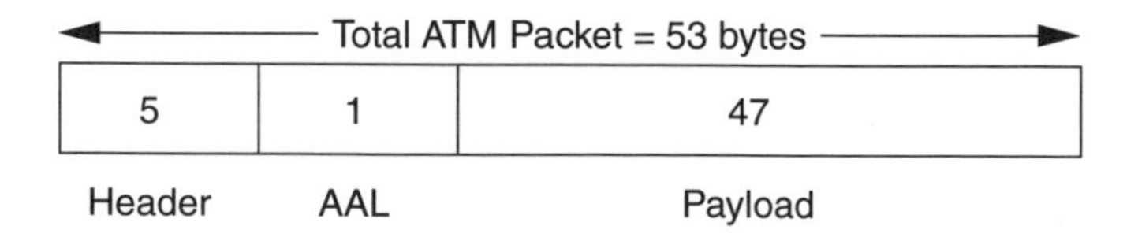

Figure 7.2 ATM Packet Structure

(AAL), and a 47-byte payload. Since 4 * 47 = 188, an MPEG 2 Transport Packet maps into four ATM packets with no overhead.

7.2.2 Transport Stream Rate

The Transport Stream defines a bitstream that is optimized for communicating or storing one or more programs of MPEG 2 coded data and other data for error-prone environments. The Transport Stream rate is defined by the values and locations of Program Clock Reference (PCR) fields.

Generally, there are separate PCR fields for each program. A single Transport Stream may contain a number of such programs, each with a different time base. Note that this is necessary for DBS because the program materials that are multiplexed into a Transport Stream for a particular transponder arrive at the uplink site in a wide variety of ways. Each DBS transponder has a fixed bandwidth, so the Transport Stream bit rate is constant for each transponder.

7.2.3 Transport Stream Packet Header

The Transport Stream Packet Header contains all of the information required by a decoder to decode the various packet structures. Figure 7.3 shows the overall structure. Referring to this figure, the Transport Stream packets can

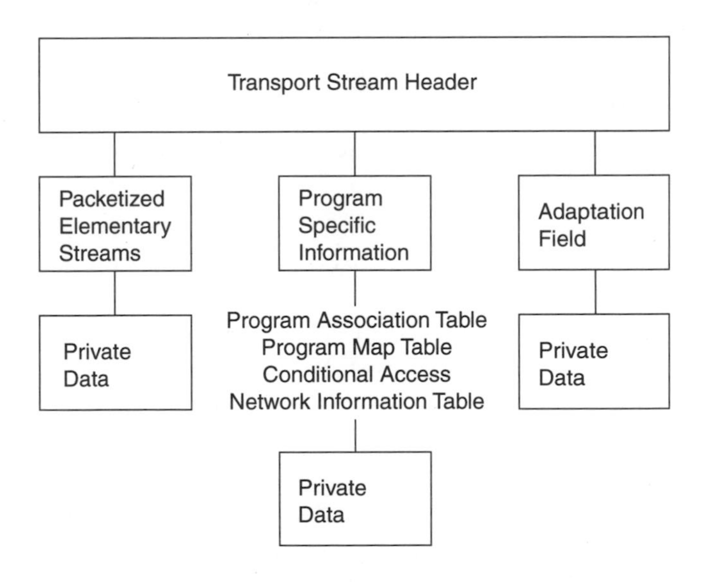

Figure 7.3 Transport Stream Header Structure

contain three types of payloads: Packetized Elementary Streams, Program Specific Information, and Adaptation fields. Only one of these packet types can be the payload at any one time (there is one exception, but it is of minor importance). The PSI packets are composed of four different types of packets. Once again, only one is present in a particular Transport Stream packet. Each of the three payload types can carry private data.

It is instructive to follow through on how a decoder can parse the Transport Stream Header. First, the **adaptation_field_control** field is checked to see if it is 2 or 3. If it is, the decoder knows that the packet contains an Adaptation field, and starts to further parse it. If the **adaptation_field_control** is 1 or 3, the decoder continues to parse the Transport Stream Header to determine whether PES or PSI data is contained in this packet. If, at this point in the decoding, the **adaptation_field_control** is not 1 or 3, the packet is discarded as not containing useful data.

Assuming then that the **adaptation_field_control** is 1 or 3, the decoder determines whether the PID value is for PES or PSI. Either the PES or PSI payload types can carry Private data. If the packet is an Adaptation field, the **transport_private_data_flag** is checked. If it is a '1', the packet is carrying Private data. If '0', the packet is carrying another type of Adaptation field.

The following list discusses the constituents of the Transport Stream Packet Header.

All packets contain sync bytes which allow the decoder to synchronize to the bitstream. The **sync_byte** value is '0100 0111' (0x47)—remember that the notation 0x means hexadecimal. Sync byte emulation in the rest of the bitstream should be avoided.

When the bitstream contains at least one uncorrectable bit error, the **transport_error_indicator** is set to '1'.

The **payload_unit_start_indicator** is a 1-bit flag that has a specific meaning for Transport Stream packets that carry PES packets or PSI data. When this flag is set (a '1'), this packet contains the first byte of a PES packet, or the first byte of a PSI section. If not set, it does not carry such a byte. When a Transport Packet contains PSI data and the **packet_unit_start_indicator** is '1', the first byte after the header is a **pointer_field.**

The **pointer_field** (PSI payload only) is an 8-bit field whose value is the number of bytes immediately following the **pointer_field** until the first byte of the first section in the payload of that packet (so a value of 0x00 in the **pointer_field** indicates that the section starts immediately after the **pointer_field**).

Packets with the **transport_priority** set to '1' indicate that the packet is of greater priority than other packets with the same PID that has this flag set to '0'.

The PID is a 13-bit field that identifies the packet payload. PID values for the Program Association Table and the Conditional Access Table are specified. Other PID values are reserved.

The **transport_scrambling_control** (a better name would be **transport_encryption_control**), 2-bit field indicates the encryption mode of the Transport Stream Packet payload. Transport Stream Packet Headers, including the Adaptation field when present, cannot be encrypted.

The **adaptation_field_control** indicates whether this Transport Stream Packet Header is followed by an Adaptation field and/or a payload.

Packets with the same PID are counted by a 4-bit counter and presented in the Transport Stream Header as the **continuity_counter.** The 4-bit counter wraps around to '0' after it reaches '1111'. The **continuity_counter** cannot be incremented when the payload is an Adaptation field.

When the **discontinuity_counter** state is true in any Transport Stream packet, the **continuity_counter** in the same packet may be discontinuous with respect to the previous Transport Stream packet of the same PID.

7.2.4 Timing Model

MPEG 2 Systems, Video, and Audio all have a timing model in which the end-to-end delay from the signal input to an encoder, to the signal output from a decoder, is a constant. This delay is the sum of all the delays from encoding to presentation. The inter-Picture interval and audio sample rate are the same at the decoder as at the encoder.

7.2.5 Conditional Access

Chapter 6 noted the importance of Conditional Access (CA) for a DBS system, so MPEG 2 Systems must support CA. Encryption for Conditional Access to programs is encoded in the Transport Stream in MPEG 2 Systems. Elements of the standard provide specific support for such systems.

7.2.6 Multiplex-wide Operations

Multiplex-wide operations include the coordination of data retrieval off the channel, clock adjustment, and buffer management. In general, for the broadcast case, Elementary Stream decoders must slave their timing to the data received from the channel to avoid overflow or underflow of decoder buffers.

7.2.7 Transport Stream System Target Decoder

A hypothetical decoder known as the Transport Stream System Target Decoder (T-STD) is used to define timing. Byte arrival and decoding event timing are set out in MPEG 2 Systems. A detailed discussion of the timing model of the T-STD is beyond the scope of this book. For more details, refer to section 2.4.2 of ISO/IEC 13818-1.

Payload (non-Adaptation field): The payload of a Transport Packet consists of data from the PES packets, PSI sections, or Private data that are labeled in the standard **data_bytes.**

7.3 Individual Stream Operations (PES Packet Layer)

Demultiplexing and synchronizing playback of multiple Elementary Streams are the principal stream-specific operations.

7.3.1 Demultiplexing

Elementary Streams are multiplexed into Transport Streams. Audio and video streams are the primary ESs. Only one Elementary Stream may be in a PES packet. PES packets are normally relatively long. Packet ID codes in the Transport Stream permit the decoding and demultiplexing required to reconstitute ESs.

7.3.2 Synchronization

Time stamps allow synchronization among multiple Elementary Streams. The System Clock Reference (SCR), the Program Clock Reference (PCR), and the optional Elementary Stream Clock Reference (ESCR) have extensions with a resolution of 27 MHz. Because presentation time stamps apply to the decoding of individual ESs, they reside in the PES packet layer of Transport Streams.

The PCR time stamp allows synchronization of a decoding system with the channel. The PCR time stamp encodes the timing of the bitstream itself and is derived from the same time base used for the audio and video. Because each program may have its own time base, there are separate PCR fields for each program in a Transport Stream that has multiple programs. A program can have only one PCR time base associated with it.

7.3.3 Relation to Compression Layer

While compressed audio and video Elementary Streams contain their own information for decoding, these streams are the payloads for PES packets. The PES packets do not concern themselves with the nature of the payload. For example, video start codes can occur in the middle of a PES packet. However, time stamps encoded in PES Packet Headers apply to presentation times of compression layer constructs (namely, presentation units).

7.3.4 PES Packets

Packetized Elementary Stream

Transport Streams for multimedia content are logically constructed from Packetized Elementary Stream packets, as indicated in Figure 7.1. Elementary Stream Clock Reference fields and Elementary Stream Rate (**ES_Rate**) fields must be included in PES packets. PES packets are the dominant form of Transport Packet payload. The structure of these packets is discussed in the following section.

PES Packet Structure

PES packets start with a Packet Header. The header for PES packets is much longer and more involved than the other headers in MPEG 2 Systems because they must provide a great deal of information regarding the constituent elementary streams. This book discusses the more important header parameters and summarizes the remainder. For full details the reader is referred to the ISO standard.

As with most packets, the first parameter is a sync or **start_code.**

The **packet_start_code_prefix** is the bit string—'0000 0000 0000 0000 0000 0001' (0x000001). As usual, care should be taken not to emulate the start code in the remainder of the packet.

The **stream_id** specifies the type and number of the Elementary Streams and may take values in the range 0xBC to 0xFF. The value 0xBC signifies the **Program_Stream_Map.** All IDs that start with 0xC and 0xD are used for audio streams. 0xE IDs are used for video streams and 0xF0 and 0xF1 IDs are used for Entitlement Control Messages (ECM) and Entitlement Management Messages (EMM).

Table 7.1 What **PTS_DTS_flags** Mean

PTS_DTS_flags	Meaning
'00'	Neither a **PTS** nor a **DTS** field is present
'01'	Forbidden
'10'	**PTSD** field is present
'11'	Both **PTS** and **DTS** fields are present

The **PES_packet_length** is a 16-bit field specifying the number of bytes in the PES packet. Thus, a PES packet could be 65,536 bytes long.

The **PES_scrambling_control** is a 2-bit field that indicates the scrambling mode of the PES packet payload. Once again, encryption control would have been a better name.

The next 11 parameters in the header are the flags: **PES_priority, data_alignment_indicator, copyright, original_or_copy, PTS_DTS_flags, ESCR_flag, ES_rate_flag, DSM_trick_mode_flag,[1] additional_copy_info_flag, PES_CRC_flag, and PES_extension_flag.** All these parameters are 1-bit flags, indicating the presence or absence of a condition, except for **PTS_DTS_flags,** which is 2 bits. Table 7.1 explains **PTS_DTS_flags.**

If the **PES_extension_flag** is '1', the **PES_header_data_length** 8-bit field specifies the total number of bytes occupied by the optional fields and any stuffing bytes contained in this PES Packet Header. The **marker_bit** is a 1-bit flag that always has the value '1'.

The next three parameters are the various time stamps (if present): **PTS (presentation_time_stamp), DTS (decoding_time_stamp),** and **ESCR (elementary_stream_clock_reference).** The **PTS** and **DTS** indicate the time for presentation and decoding. They are 33-bit numbers that are measured in the number of 27-MHz clock periods divided by 300. The **ESCR** is a 42-bit value comprised of a 33-bit base (**ESCR_base**) and a 9-bit extension (**ESCR_ext**). The base is measured in periods of a 90-kHz clock.

The **ES_rate (elementary_stream_rate)** field is a positive integer specifying the PES stream rate.

[1] In MPEG a "trick mode" is a mode usually associated with a digital storage medium. Thus, fast forward, slow motion, rewind, and so on, are trick modes.

trick_mode_control is a 3-bit field that indicates which trick mode is applied to the associated video stream. Since trick modes are not applicable for DBS, they are not discussed further here.

field_id is a 2-bit field that indicates which field(s) are to be displayed in trick modes. This is not relevant for DBS.

In MPEG 2 Video, a Macroblock is a 16-by-16 pixel Block (Y axis). If some Macroblocks may be missing, **intra_slice_refresh** is set to '1'. This knowledge can be used for error concealment.

The 2-bit **frequency_truncation** field can indicate that a restricted set of coefficients may have been used in coding the video data.

field_rep_cntrl is a 5-bit field that indicates certain display parameters in Video trick modes. Not applicable for DBS.

The **additional_copy_info** is a 7-bit field that contains Private data relating to copyright information.

The **previous_PES_packet_CRC** is a 16-bit CRC (see Appendix E).

There then follow five 1-bit flags: **PES_private_data_flag, packet_header_field_flag, program_packet_sequence_counter_flag, P-STD_buffer_flag,** and **PES_extension_flag_2.** These flags indicate the presence or absence of the indicated condition.

Private data in a PES packet is carried in **PES_private_data,** a 16-bit field. This data must not emulate the **packet_start_code_prefix** (0x000001).

The length, in bytes, of the **packet_header_field** is indicated by the 8-bit field **packet_field_length.**

The PES packets of a particular program are counted by the 7-bit **program_packet_sequence_counter.**

The remainder of the header parameters—**MPEG1_MPEG2_identifier, original_stuff_length, stuffing_byte, P-STD_buffer_scale** [Program Stream only], and **PES_extension_field_length**—are not important for DBS.

PES Data Packets

The **PES_packet_data_bytes** are contiguous bytes of data from the ES indicated by the packet's **stream_id** or **PID.** The **padding_byte** is a fixed 8-bit value equal to '1111 1111'. The decoder discards it.

7.4 Program Specific Information

Program Specific Information includes both MPEG 2 Systems required data and Private data that enable demultiplexing of programs by decoders. Programs or Elementary Streams, or parts of them, may be conditionally accessed. However, Program Specific Information *cannot* be encrypted.

In Transport Streams, PSI consists of four table structures. The tables that comprise PSI packets have a very similar structure, as shown in Table 7.2. Common fields are in bold. Fields unique to a particular table are in regular type.

7.4.1 Common Field Types

Because a number of field types are common to all four tables,[2] they will be described first. Then, the fields unique to each table are discussed.

The **table_id** is an 8-bit field that identifies the content of a Transport Stream PSI section, as follows: 0x00 defines the **program_association_section;** 0x01 defines the **conditional_access _section;** 0x03 to 0x3F are reserved; 0x40 to 0xFE are User Private; and 0xFF is forbidden.

The **section_syntax_indicator** is a 1-bit field that must be set to '1'.

The **section_length** is a 12-bit field, the first two bits of which are '00'. It specifies the number of bytes of the section, starting immediately following it, including the CRC.

The **version_number** is a 5-bit field that is the version number of the Program Association Table. Whenever the definition of the Program Association Table changes, the version number is incremented by one. When the value 31 is reached, it wraps around to '0'. The **current_next_indicator** determines whether the **version_number** is that of the currently applicable Program Association Table or that of the next applicable PAT.

The **current_next_indicator** is a 1-bit field that, when set to '1', indicates that the PAT sent is currently applicable. When the bit is set to '0', it indicates that the table sent is not yet applicable and will be the next table to become valid.

---

[2] In discussing common fields, the description uses the language for the Program Association Table (PAT). In this common section, it applies to all of the relevant tables.

Table 7.2 PSI Table Parameters with Common Items Indicated (in Bold)

Program Association Table	Program Map Table	Conditional Access Table	Private Section
table__id	**table_id**	**table_id**	**table_id**
section_syntax_indicator	**section_syntax_indicator**	**section_syntax_indicator**	**section_syntax_indicator**
section_length	**section_length**	**section_length**	
transport_stream_id	program_number		private_indicator
version_number	**version_number**	**version_number**	* **version_number**
current_next_indicator	**current_next_indicator**	**current_next_indicator**	* **current_next_indicator**
section_number	**section_number**	**section_number**	* **section_number**
last_section_number	**last_section_number**	**last_section_number**	* **last_section_number**
program_number	PCR_PID		* private_section_length
network_PID	program_info_length		* private_data_byte
program_map_PID	stream_type		* table_id_extension
	elementary_PID		
	ES_info_length		
CRC_32	**CRC_32**	**CRC_32**	**CRC_32**

The **section_number** is an 8-bit field that gives the number of the current section. The **last_section_number** is an 8-bit field that specifies the number of the last section (the section with the highest **section_number**) of the complete Program Association Table.

CRC_32—This field is 32-bit with various CRC values (see appendix E).

7.4.2 Program Association Table

All Transport Streams contain one or more Transport Stream packets with a PID value of 0x0000. These Transport Stream packets together contain a complete list of all programs within the Transport Stream. Any changes in the programs carried within the Transport Stream are described in an updated version of the Program Association Table carried in Transport Stream packets with PID value 0x0000. These sections use **table_id** value 0x00. Only sections with this value of **table_id** are permitted within Transport Stream packets with a PID value of 0x0000.

The maximum number of bytes in a section of an MPEG 2 Systems-defined PSI table is 1K (1,024 bytes). The maximum number of bytes in a **private_section** is 4K (4,096 bytes). There are no restrictions on the occurrence of start codes, sync bytes, or other bit patterns in PSI data, whether MPEG 2 Systems or Private.

The Program Association Table provides the correspondence between a **program_number** and the PID value of the Transport Stream Packets that carry the program definition. The **program_number** is the numeric label associated with a program. Program number 0x0000 is reserved to specify the network PID. This identifies the Transport Stream packets that carry the Network Information Table.

Definition of Unique Fields in PAT Section

A Transport Stream can be separated from any other multiplex within a network by the **transport_stream_id,** a 16-bit field.

The program to which the **program_map_PID** is applicable is set by the 16-bit **program_number.** If this is set to 0x0000, then the following PID reference must be the network PID. The **program_number** may be used as a designation for a broadcast channel, for example.

The **network_PID** (optional) is a 13-bit field that specifies the Network Information Table.

The Transport Stream packets that must contain the **program_map_section** applicable for a program is specified by the 13-bit **program_map_PID** field.

7.4.3 Program Map Table

All Transport Streams must contain one or more Transport Stream packets with PID values that are labeled under the Program Association Table as Transport Stream packets containing Program Map sections. Every program listed in the PAT is described in a single Transport Stream Program Map section.

The most recently transmitted version of the **TS_program_map_section** with the **current_next_indicator** set to a value of '1' always applies to the current data within the Transport Stream. Any changes in the definition of any of the programs carried within the Transport Stream are described in an updated version of the corresponding section of the Program Map Table carried in Transport Stream packets with the PID value identified as the **program_map_PID** for that specific program. During the continuous existence of a program, including all of its associated events, the **program_map_PID** cannot change.

Sections with a **table_id** value of 0x02 contain Program Map Table information. Such sections may be carried in Transport Stream packets with different PID values.

A *program definition* is a mapping between a program number and the Elementary Streams that comprise it. The Program Map Table provides the complete collection of all program definitions for a Transport Stream. The table may be segmented into one or more sections, before insertion into Transport Stream packets.

Definition of Unique Fields in Transport Stream Program Map Section

program_number (see Program Association Table description).

The PCR fields valid for the program specified by **program_number** must be identified by a 13-bit **PCR_PID.**

The number of bytes of the descriptors are indicated by the 12-bit **program_ info_length,**[3] the first two bits of which must be '00'.

The type of ES or payload specified by the **elementary_PID** is indicated by the 8-bit **stream_type** field.

The PID of the Transport Stream packets that carry the associated Elementary Stream or payload is indicated by the 13-bit **elementary_PID.**

Elementary Stream Descriptors

Elementary Stream Descriptors provide additional information about the ESs. All stream descriptors have a format that begins with an 8-bit tag value. The tag value is followed by an 8-bit descriptor length and data field.

Stream Descriptors

The 8-bit fields—**descriptor_tag,** which identifies each descriptor, and **descriptor_length,** which specifies the number of bytes of the descriptor immediately following the **descriptor_length** field—are common to all stream descriptors.

Although the 8-bit structure of the stream tag would allow for 256 descriptors, only 13 are used. Tag values 2 (**video_stream_descriptor**), 7 (**target_ background_grid_descriptor**), and 8 (**video_window_descriptor**) all describe attributes of Video streams.

---

[3]For information on the **ES_info_length** field, see **program_info_length.**

Video Stream Descriptors

The **video_stream_descriptor** identifies key parameters of the video ES. In particular, the frame rate code is identified. The profile and level are also indicated (see Chapter 8), along with other parameters.

The **target_background_grid_descriptor** describes how to place the video display when the display area is larger than the video being transported.

The **video_window_descriptor** is used to describe the window characteristics of the associated ES. Its values reference the **target_background_grid_descriptor** for the same stream.

Audio Stream Descriptor (tag value 3)

The **audio_stream_descriptor** provides basic information that identifies the coding version of an audio ES.

Registration Descriptor (tag value 5)

The **registration_descriptor** provides a method to uniquely and unambiguously identify formats of private data. It includes a 32-bit value obtained from a registration authority as designated by ISO.

Conditional Access Descriptor (tag value 9)

If any systemwide Conditional Access management information exists within a Transport Stream, a **CA_descriptor** must be present in the appropriate map section. This descriptor is used to specify both systemwide CA management information—EMMs and ES-specific information such as ECMs.

When the **CA_descriptor** is in the **TS_program_map_section** (**table_id** = 0x02), the **CA_PID** points to packets containing program-related access control information such as ECMs. When the **CA_descriptor** is found in the **CA_section** (**table_id** = 0x01), the **CA_PID** points to packets containing systemwide and/or access control management information such as EMMs.

ISO 639 Language Descriptor (tag value 10)

ISO 639 Part 2 defines a 3-character code for languages. The **ISO_639_language_descriptor** can be used to specify the language of the associated ES.

System Clock Descriptor (tag value 11)

This descriptor conveys information about the system clock that was used to generate the time stamps. The clock frequency accuracy is

$$\textbf{clock\_accuracy\_integer} * 10^{-\textbf{clock\_accuracy\_exponent}} \text{ ppm}$$

where **clock_accuracy_integer** is 6 bits and the **clock_accuracy_exponent** is 3 bits.

If an external clock reference was used, the **external_clock_reference_indicator** is set. If the system clock is more accurate than the 20-ppm accuracy required, then the accuracy of the clock can be communicated by encoding it in the **clock_accuracy** fields.

If **clock_accuracy_integer** = 0, the system clock accuracy is 20 ppm. When both parts of the descriptor are used, the clock accuracy pertains to the external reference clock.

Copyright Descriptor (tag value 13)

The **copyright_descriptor** contains a **copyright_identifier** that is a 32-bit value obtained from a Registration Authority as designated by ISO.

Maximum Bitrate Descriptor (tag value 14)

The **maximum_bitrate_descriptor** is coded as a 22-bit positive integer in the field expressing the maximum bit rate in units of 50 bytes per second.

7.4.4 Conditional Access Table

Transport Stream packets must convey the appropriate **CA_descriptors** whenever one or more ESs within a Transport Stream are encrypted. Transport Stream packets with a PID value 0x0001 contain these descriptors. A complete version of the Conditional Access Table must be maintained in the decoder. When the **current_next_indicator** is '1', that table is applied to the encrypted files in the Transport Stream. Any changes in encryption making the existing ISO table invalid or incomplete are described in an updated version of the Conditional Access Table. All these sections use a **table_id** value of 0x01. Only sections with this **table_id** value are permitted within Transport Stream packets with a PID value of 0x0001.

The Conditional Access Table provides the association between one or more CA systems, their EMM streams, and any special parameters associated with them. Conditional Access fields only use the common fields.

7.4.5 Network Information Table

The Network Information Table is optional and its contents are Private. If present, it is carried within Transport Stream packets that have the same PID value, called the **network_PID.** The **network_PID** value is defined by the user and, when present, is found in the Program Association Table under the reserved **program_number** 0x0000. If the Network Information Table exists, it takes the form of one or more **private_sections.**

7.4.6 Syntax of the Private Section

In addition to defined PSI tables, it is possible to carry Private data tables. These may be structured in the same manner used for defined PSI tables so that the syntax for mapping this Private data is identical to that used for the mapping of a defined PSI table. For this purpose, a Private section is defined. If the Private data is carried in Transport Stream packets with the same PID value as Transport Stream packets carrying Program Map Tables (as identified in the PAT), then the **private_section** syntax and semantics are used.

A Private table may be made of several **private_sections**, all with the same **table_id**.

7.5 Adaptation Field

Recall that the Adaptation field is signaled by a 2-bit flag in the Transport Packet Header. By sending infrequently changing parameters in this manner, rich collateral information can be conveyed without substantially reducing overall header efficiency.

The first parameter in the Adaptation field is the **adaptation_field_length,** an 8-bit field specifying the number of bytes in the Adaptation field immediately following it.

The **discontinuity_indicator** is a 1-bit field that, when set to '1,' indicates that the discontinuity state is true and that the PCR, which is a sample of the **system_clock_frequency** for that program, is about to change. When the discontinuity state is true for a Transport Stream packet that contains a **PCR_PID,** the next PCR in a Transport Stream packet with the same PID represents the new clock frequency.

These are the 8 1-bit flags that follow: **discontinuity_indicator, random_access_indicator, elementary_stream_priority_indicator, PCR_flag, OPCR_flag, splicing_point_flag, transport_private_data_flag,** and **adaptation_field_extension_flag**. As usual with such flags, they indicate the presence or absence of a particular condition.

The **discontinuity_indicator** being set means the PCR is going to be updated.

The **random_access_indicator** being set means that the next PES packet contains either the first byte of a video sequence header or the first byte of an audio frame.

When the **PCR_flag** is set, a new, two-part PCR is contained in this Adaptation field. This PCR is a 42-bit field coded in two parts.

program_clock_reference_base is a 33-bit field in units of 1/300 multiplied by the system clock frequency (90 kHz).

The **program_clock_reference_extension** is a 9-bit field in units of system clock frequency (27 MHz).

The following are not important for DBS: **original_program_clock_reference_base** and **original_program_clock_reference_extension.**

splice_countdown is not important for DBS.

The **private_data_length** is an 8-bit field specifying the number of **private_data** bytes immediately following it.

The **private_data_byte** is an 8-bit field that shall not be specified by ISO/IEC.

The **adaptation_field_extension_length** is an 8-bit field that indicates the length of the extended Adaptation field data that follows it.

There then follow 4 more 1-bit flags: **ltw_flag (legal_time_window_flag), piecewise_rate_flag, seamless_splice_flag,** and **ltw_valid_flag (legal_time_ window_valid_flag).** As usual, these flags indicate the presence or absence of a condition.

ltw_offset (legal_time_window_offset) is a 15-bit field that is part of T-STD timing.

piecewise_rate is a 22-bit field that is a positive integer specifying the bit rate over all transport packets of this PID, starting with the packet containing this field and continuing until the next occurrence of this field. The value of the **piecewise_rate** is measured in units of 50 bytes per second. A value of '0' is forbidden.

splice_type is not important for DBS.

The **DTS_next_au (decoding_time_stamp_next_access_unit)** is a 33-bit field that indicates the value of the DTS of the next access unit. The value is derived using the system time clock valid before the splice occurred.

stuffing_byte is a fixed 8-bit value equal to 0xFF that can be inserted by the encoder. It is discarded by the decoder.

CHAPTER EIGHT

MPEG 2 Video Compression

Suppose video is represented by frames of 480 pixels vertical by 720 pixels horizontal (the approximate size of CCIR 601-1). These are arrays of numbers that range from 0 to 255, with the value representing the intensity of the image at that point. This video is to be displayed at 30 frames per second.

There are three color planes (and hence three arrays). The bit rate for this video can thus be calculated as:

pixels per frame per color plane (480 * 720):	345,600
times three color planes:	1,036,800
times eight bits per pixel:	8,294,400 bits per frame
times 30 frames per second:	249 Mbps

8.1 The Need for Video Compression

Most studies have shown that the economics of a DBS system require from four to eight TV channels per transponder. If the information bit rate per transponder is 30 Mbps, the total bit rate (video, audio, and other data) must be 3.75 Mbps to 7.5 Mbps (on average) per audio and/or video service. The audio and data will require approximately 0.2 Mbps per service. Thus, the video must be between 3.55 Mbps and 7.3 Mbps. Comparing this to the uncompressed 249 Mbps, it is clear that the video has to be compressed by 34 to 70 to 1 in order to have an economically viable DBS system. This chapter explains how this is achieved.

8.2 Profiles and Levels

All MPEG standards are generated as *generic*. This means that they are intended to provide compression for a wide variety of applications. The MPEG 2 Video standard lists a large number of possible applications, including DBS.

The extremely wide range of applications requires a commensurately large range of bit rates, resolutions, and video quality. To cope with this wide range of parameters, MPEG Video uses the concept of Profiles and Levels.

A *Profile* is a defined subset of the entire bitstream syntax. Within the bounds of a particular profile, there is a wide range of permissible parameters. *Levels* indicate this range. The Profiles and Levels are orthogonal axes. Section 8 of ISO 13818-2 has a number of tables that define the Profile and Level parameters.

For example, the Maximum Bit Rate table shows that the maximum bit rate for the Main Profile can range from 15 Mbps for the Main Level to 80 Mbps for the High Level. While there are five Profiles and four Levels, only 11 of the combinations have entries in the ISO table. The other nine entries are not defined. The following are the defined Profiles and Levels:

Profile: Simple
　　Level: Main

Profile: Main
　　Levels: Main, High 1,440, High

Profile: SNR
　　Level: Low, Main

Profile: Spatial
　　Level: High 1,440

Profile: High
　　Level: Main, High 1,440, High

These Profiles and Levels cover an extremely wide range of video parameters; however, all known DBS systems currently use Main Profile at Main Level (written MP@ML). Thus, although High Profile may be used for DBS in the future, except for Chapter 13 where we address future services, the remainder of this book covers MP@ML.

8.3 Digital Video Primer

Digital video is a sequence of frames in which the individual frames are considered to be samples of an analog image on a rectangular sampling grid. The individual samples are called picture elements, or pixels. This is a very good model when the acquisition device is a CCD camera; when film is digitized; or when an analog camera's output is sampled, held, and then converted to a digital format by an analog-to-digital converter.

8.3.1 Color

To produce a color image, three color axes are required. Most acquisition and display devices use the additive primaries—red, green, and blue (RGB)—to achieve this. However, the RGB axes are correlated with each other. Thus, virtually all compression algorithms perform a preprocessing step of color coordinate conversion that creates a luminance component (Y) and two chrominance components (Cr and Cb). The Y component displayed by itself is the black-and-white video. At the decoder output, a postprocessing step converts the color space back to RGB for display.

The color coordinate conversion consists of multiplying a three-component vector (RGB) by a 3-by-3 matrix.

$$\begin{bmatrix} Y \\ C_r \\ C_b \end{bmatrix} = \begin{bmatrix} .299 & .587 & .114 \\ -.169 & -.331 & .500 \\ .500 & -.419 & -.081 \end{bmatrix} * \begin{bmatrix} R \\ G \\ B \end{bmatrix}$$

At the output of the decoder, a post-processing step converts the [Y, Cr, Cb] vector back to [R, G, B] by the following matrix/vector multiplication.

$$\begin{bmatrix} R \\ G \\ B \end{bmatrix} = \begin{bmatrix} 1.000 & 0 & 1.404 \\ 1.000 & -.3434 & -.712 \\ 1.000 & 1.773 & 0 \end{bmatrix} * \begin{bmatrix} Y \\ C_r \\ C_b \end{bmatrix}$$

8.3.2 Interlace

In the 1930s, when TV was being developed, the cathode ray tube (CRT) was the only choice available as a display device. At 30 frames per second,[1] the flicker on the CRT was unacceptable. To solve this, the TV developers invented the concept of interlace.

Interlaced frames are separated into two fields. Temporally, the rate of the fields is 60 fields per second, twice the frame rate. This refresh rate makes

[1] The NTSC values are used here. For PAL or SECAM, the frame rate is 25 frames per second and the field rate is 50 fields per second.

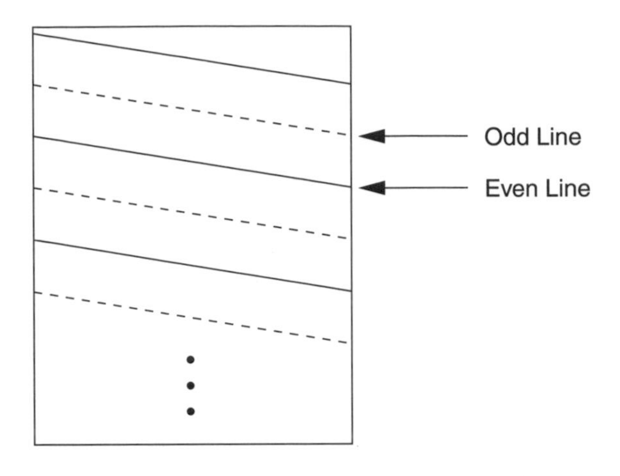

Figure 8.1 Interlace Frame

the flicker acceptable (actually, some flicker can be seen at 60 fields per second; at 72 fields per second, it disappears entirely).

Spatially, the horizontal lines from the two fields are interleaved as shown in Figure 8.1. It is frequently claimed that interlace provides a bandwidth compression of 2:1. This is nonsense. While the bandwidth required is half of what would be required if the frames were sent at the field rate, only half the information is sent with each field.

A display that presents the frame lines sequentially is called *progressive.* At this point in time, technology has developed to where progressive display is economically feasible, and it would significantly improve image quality. However, virtually all of the 2 to 3 billion TV sets in the world use interlace, so any DBS system must be able to accommodate interlacing.

8.3.3 Chroma Subsampling

To achieve the best quality, digital video should be sampled on a grid with equal numbers of samples for each of the three color planes. However, TV engineers discovered that the Human Vision System (HVS) is less sensitive to resolution in the Cr and Cb axes (the color axes) than on the Y, or luminance, axis. It is not surprising that this lesser sensitivity to Cr and Cb was taken advantage of, given the huge amounts of data generated by digital video.

The CCIR 601-1 Standard for recording digital video uses "4:2:2" chroma subsampling. In digital terms, this means that the Cr and Cb chroma planes are decimated horizontally by a factor of two relative to the luma. If the color planes at origination are 480-by-720 pixels, after this decimation

Table 8.1 Position of Luminance and Chrominance Samples 4:2:2

⊗	X	⊗	X	⊗	X	⊗	X
⊗	X	⊗	X	⊗	X	⊗	X
⊗	X	⊗	X	⊗	X	⊗	X
⊗	X	⊗	X	⊗	X	⊗	X
⊗	X	⊗	X	⊗	X	⊗	X
⊗	X	⊗	X	⊗	X	⊗	X

the luminance plane is still 480-by-720 pixels, but the Cr and Cb axes are each 480-by-360 pixels. Another way of stating this is that the bits per pixel have been reduced from 24 (8 for each color plane) to 16, a *compression* of one-third. Table 8.1 shows the relative location of the luminance (X) and chrominance (O) samples for 4:2:2 sampling. The circled X ⊗ is for both Cr and Cb samples.

The developers of the MPEG standard took another step. Because the decimation of Cr and Cb in the horizontal direction is asymmetric, they also explored decimation in the vertical direction. This led to the "4:2:0" sampling structure that decimates the Cr and Cb color planes vertically as well as horizontally. Thus, while the Y plane remains at 480-by-720 pixels, the Cr and Cb color planes are 240-by-360 pixels. Table 8.2 shows the position of the luminance and chrominance samples for 4:2:0. The O represents both Cr and Cb samples.

The additional decimation vertically reduces the aggregate bits per pixel to 12. Thus, the 4:2:0 sampling has produced an initial compression of 2:1.

Table 8.2 Position of Luminance and Chrominance Samples 4:2:0

X	X	X	X	X	X	X	X
O		O		O		O	
X	X	X	X	X	X	X	X
X	X	X	X	X	X	X	X
O		O		O		O	
X	X	X	X	X	X	X	X
X	X	X	X	X	X	X	X
O		O		O		O	
X	X	X	X	X	X	X	X

8.4 Structure of MPEG 2 Coded Video

The structure of the coded video data has a nested hierarchy of headers and the following coded data, which can comprise six levels:

Video Sequence
    Group of Pictures (optional)
        Picture(s)
            Slice
                Macroblock
                    Block

When the decoding of the MPEG 2 Video bitstream is discussed in section 8.6, it will be shown that the problem is to start with a bitstream and parse it down to an 8-by-8 Block, which can then be recreated for display.

The beginning of each of the levels in the hierarchy is indicated by a unique 32-bit (eight-hex character) start code. The first six hex characters are common to all of the codes: 0x000001. A two-hex character suffix, which is unique to the particular level, is then added. In the rest of this chapter only the two-hex character codes are used—the **picture_start_code** = 0x00. The **slice_start_code** ranges from 0x01 to 0xAF. These sequentially indicate the Slices within a picture. The remainder of the codes in MPEG 2 Video are of the form 0xBX, where X ranges from 0 to 8. Three of these, 0xB0, 0xB1, and 0xB6, are reserved. Of the remaining six, three involve the Sequence level—**sequence_header_code** (0xB3), **sequence_ error_code** (0xB4), and **sequence_ end_code** (0xB7). The remaining three are: **user_data_start_code** (0xB2), **extension_start_code** (0xB5), and **group_start_ code** (0xB8).

8.4.1 Video Sequence

The Video Sequence is the highest-level structure in the hierarchy. It begins with a sequence header, **sequence_header_code** (0xB3), that may then be followed by a Group of Pictures (GOP) header and one or more coded frames. The order of transmission of the frames is the order in which the decoder processes them, but not necessarily the correct order for display.

After the **sequence_header_code,** the sequence header contains information about the video, which usually won't change too often. This includes information about the horizontal and vertical sizes (**horizontal_size_value** and **vertical_size_value**) that are unsigned 12-bit integers. Thus, without extensions, a 4,096-by-4096 pixel image size can be specified.

The aspect ratio in Video is the ratio of the number of horizontal pixels to the number of vertical pixels in a frame. The parameter **aspect_ratio_information,** which contains 4 bits, provides this information. The value '0000' is forbidden. A square picture (1:1 aspect ratio) has the code '0001'. The standard TV aspect ratio (4:3) has the code '0010'. The 16:9 ratio code is '0011' and a 2.21:1 ratio code is '0100'.

The **frame_rate_code** is another 4-bit unsigned integer that indicates the frame rate. The code '0000' is forbidden. The values '1001' through '1111' are reserved. Codes '0001' to '1000' indicate 23.976, 24, 25, 29.97, 30, 50, 59.94, and 60 frames per second.

The bit rate is a 30-bit integer. It indicates the bit rate in multiples of 400 bps. The lower 18 bits are indicated by **bit_rate_value.** The upper 12 bits are in **bit_rate_extension,** which is contained in **sequence_extension.**

The next parameter in the **sequence_header** is a single bit, which is always a '1', called the **marker_bit.** It is included to prevent **start_code** emulation.

Next comes the **vbv_buffer_size_value.** The parameter **vbv_buffer_size** is an 18-bit integer that defines the size of the buffer in the decoder. The first 10 bits of this value are given by **vbv_buffer_size_value.** The value of the buffer size is B = 16,384 * **vbv_buffer_size** bits. The upper 8 bits are in **vbv_buffer_size_extension,** which is in **sequence_extension.**

To facilitate the use of simple decoders in MPEG 1, a set of constrained parameters are defined. For MPEG 2 Video, this doesn't apply, so the 1-bit flag, **constrained_parameters_flag,** is always set to '0'.

The last parameters of the **sequence_header** permit the sending of quantization matrices. If an Intra quantizer matrix is to be sent, the 1-bit flag, **load_intra_quantizer_matrix,** is set to '1'. There then follow 64 8-bit values for **intra_quantizer_matrix[64].** A similar option is available for non-Intra quantizer matrices.

In a normal Video sequence, the **sequence_header** is followed by a **sequence_extension** that has the **start_code** 0x000001B5. The next item in the **sequence_extension** is a 4-bit **extension_start_code_identifier,** which indicates the type of extension. The values '0000', '0100', '0110', and '1011' through '1111' are reserved. Three of the **extension_start_code_identifier** codes involve the Sequence Extension—Sequence Extension ID ('0001'), Sequence Display Extension ID ('0010'), and Sequence Scalable Extension ID ('0101').

Four of the **extension_start_code_identifier** codes involve Picture coding: Picture Display Extension ID ('0111'), Picture Coding Extension ID ('1000'), Picture Spatial Scalable Extension ID ('1000'), and Picture Temporal Scalable Extension ID ('1000'). The final **extension_start_code_identifier** code is the Quant Matrix Extension ID ('0011'). If the **extension_start_code_**

identifier indicates that the extension is a regular sequence extension, the following are the parameters.

Since DBS uses only MP@ML, the parameter **profile_and-level_indication** has the value 0x48. The 1-bit flag, **progressive_sequence,** is set to '1' if the display mode is progressive. The parameter **chroma_format** is a 2-bit code that indicates whether the chroma sampling is 4:2:0 ('01'), 4:2:2 ('10'), or 4:4:4 ('11'). The value '00' is reserved.

The **horizontal_size_extension** and the **vertical_size_extension** are 2-bit unsigned integers that can be prefixes to **horizontal_size_value** and **vertical_size_value,** increasing the horizontal and vertical sizes up to 14 bits. The most significant part of the bit rate is provided by the 12-bit **bit_rate_extension.** Again, a **marker_bit** that is always '1' is inserted to prevent **start_code** emulation. The eight most significant bits of the buffer size are provided by **vbv_buffer_size_extension.**

In some applications where delay is important, no B Pictures are used. This is indicated by the 1-bit flag, **low_delay,** being set to '1'. For non-standard frame rates, the actual frame rate can be calculated from the parameters **frame_rate_extension_n** (2 bits) and **frame_rate_extension_d** (5 bits) along with **frame_rate_value** from the **sequence_header.** The calculation is **frame_rate = frame_rate_value * (frame_rate_extension_n + frame_rate_extension_d).**

The Sequence Display Extension contains information as to which Video format is used (NTSC ('010'), PAL ('001'), SECAM ('011'), and so on) in the parameter **video_format.** The 1-bit flag **colour_description** indicates if there is a change in the three 8-bit value parameters—**colour_primaries, transfer_characteristics,** and **matrix_coefficients.** Since these are detailed descriptions of the color, they won't be discussed further.

The 14-bit unsigned integer **display_horizontal_size** permits designating the horizontal size of the display. The inevitable **marker_bit,** always '1', prevents **start_code** emulation. Finally, the 14-bit unsigned integer, **display_vertical_size,** permits designating the display's vertical size.

The Sequence Scalable Extension is not used in DBS, so it is not discussed here.

The Video Sequence is terminated by a **sequence_end_code.** At various points in the Video Sequence, a coded frame may be preceded by a repeat sequence header, a Group of Pictures header, or both.

8.4.2 Group of Pictures

The GOP is a set of pictures, which includes frames that do not require information from any other frames that occur before or after (defined as Intra, or

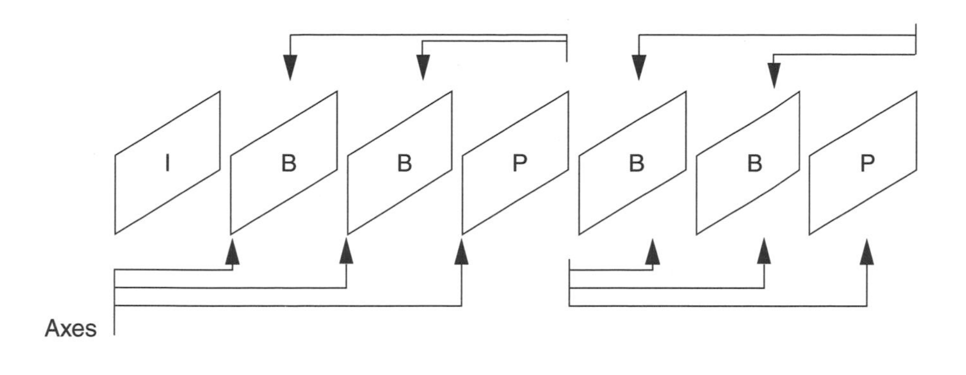

Figure 8.2a Structure of Intraframe Coding

I frames), frames that are unidirectionally predicted from a prior frame (defined as P frames), and frames that are bidirectionally predicted from prior and later frames (defined as B frames). Figure 8.2a shows a typical GOP. The header for a GOP is optional.

It is natural to ask why different types of frames are needed. Figure 8.2b shows a typical frame (it could be I, P, or B). To achieve the maximum compression, redundancies must be removed from three axes: two spatial and one temporal. The P and B frames are used to achieve temporal compression. Studies have shown that P frames require only 40% as many bits and B frames require only 10% as many bits as those needed for I frames.

The structure shown in Figure 8.2a is usually called a hybrid coder. This is because spatial compression is achieved by transform techniques and temporal compression by motion compensation. There is a feature of MPEG

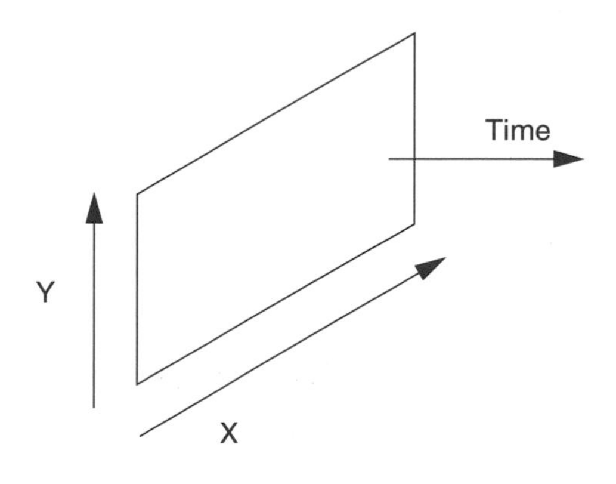

Figure 8.2b Remove Redundancies on Three Axes

compression implied by Figure 8.2a that is difficult to understand for those seeing it for the first time.

Because B frames must be created from prior I or P frames and subsequent I or P frames, the decoder must have both before the B frames can be decoded. Thus, the order of transmission cannot be the IBBPBBP . . . for both acquisition and display. Rather, the order must be IPBBPBBP. . . . Note that this puts restrictions on the encoder because it must store the frames that are going to become B frames until the frame that is going to be the following I or P frame is input into the encoder. It also increases the *latency*—the time from when encoding starts for a frame until display starts for that frame.

When the bitstream includes a Group of Pictures header, the following parameters are included. The **group_start_code** is 0xB8. A 25-bit binary string indicates the **time_code** in hours (5 bits), minutes (6 bits), and seconds (6 bits). Since there can be up to 60 pictures per second, there is a 6-bit parameter **time_code_pictures,** which counts the pictures. The **time_code** starts with a 1-bit flag, **drop_frame_flag,** which is '1' only if the frame rate is 29.97 frames per second. A **marker_bit,** always '1', is located between **time_code_minutes** and **time_code_seconds.** The 1-bit flags, **closed_gop** and **broken_link,** involve treatment of B frames following the first I frame after the GOP header.

8.4.3 Picture (Frame)

Pictures are the basic unit that is received as input and is output for display. As noted before, coded pictures can be I, B, or P. Pictures also can be either Field Pictures or Frame Pictures. Field Pictures appear in pairs: one top field picture and one bottom field picture that together constitute a frame. When coding interlaced pictures using frame pictures, the two fields are interleaved with each other and coded as a single-frame picture.

The **picture_header** starts with the **picture_start_code** 0x00. The next parameter is the **temporal_reference,** a 10-bit unsigned integer that counts the pictures and wraps around to 0 after reaching 1,023. The next parameter indicates the type of coding, **intra_coded** ('001'), **predictive_coded** ('010'), or **bidirectional_predictive_coded** ('011'). The 16-bit unsigned integer, **vbv_delay,** provides information on the occupancy of the decoder buffer. There are four parameters—**full_pel_forward_vector** (1 bit), **forward_f_code** (3 bits), **full_pel_backward_vector** (1 bit), and **backward_f_code** (3 bits)—that are not used in MPEG 2 Video. The 1-bit parameter **extra_bit_picture** is always set to '0'.

Pictures are subdivided into Slices, Macroblocks, and Blocks, each of which has its own header.

Slice

A Macroblock is comprised of 16-by-16 pixels. A Slice is a row or partial row of Macroblocks. Dividing Pictures into Slices is one of the innovations of MPEG compression. If data becomes corrupted in transmission, it may be confined to a single Slice, in which case the decoder can discard just that Slice but not the whole Picture. The Slice header also contains a quantizer scale factor, which permits the decoder to adjust the dequantization.

Figure 8.3 shows the general structure of a Slice. Note that it is not necessary for Slices to cover the entire Picture. Also, Slices may not overlap and the location of the Slices may change from Picture to Picture.

The first and last Macroblock of a Slice must be on the same horizontal row of Macroblocks. Slices also must appear in the bitstream in the order they appear in the Picture (in normal raster order, that is, left to right and top to bottom). Under certain circumstances a restricted Slice structure must be used. In this case, each Slice is a complete row of Macroblocks and the Slices cover the entire Picture.

The **slice_header** has the start codes 0x01 through 0xAF. The **slice_vertical_position** is the last 8 bits of the **slice_start_code.** For very large pictures, with more than 2,800 lines, there is a parameter **slice_vertical_position_extension,** which gives the Slice location. Because this does not apply to DBS, the calculation for it will not be given.

The parameter **priority_breakpoint** is a 7-bit unsigned integer that is used only in scalable modes and hence, not in DBS. The scalar quantizer is a 5-bit parameter with values ranging from 1 to 31. It stays the same until another value occurs in a Slice or Macroblock. The 1-bit **intra_slice_flag** is set to '1' if there are intra-Slice and reserved bits in the bitstream. If any of the

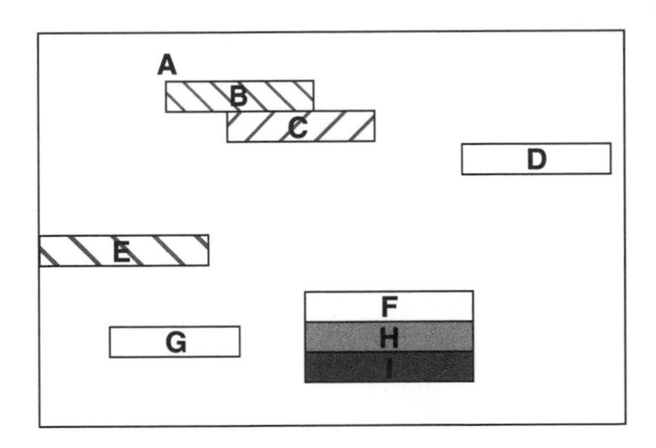

Figure 8.3 General Slice Structure

Macroblocks in the Slice are non-Intra Macroblocks, the **intra_slice** is set to '0', otherwise, it is '1'.

There are 7 bits of unsigned integer that are reserved, logically enough **reserved_bits.** The next bit is **extra_bit_slice.** If it is '1', 8 unsigned integer bits follow. This is a provision for future expansion. If **extra_bit_slice** is '0', the coding of Macroblocks within the Slice proceeds.

Macroblock

As noted earlier, a luminance Macroblock is a 16-by-16 pixel region. The chrominance part of a Macroblock depends on the chrominance sampling. Since only MP@ML is being considered in this book, the chrominance sampling is 4:2:0 and the Macroblock structure is that shown in Figure 8.4. Note that the Macroblock for 4:2:0 coding is comprised of six blocks. The luminance component has four 8-by-8 blocks and the chrominance components have one 8-by-8 block each.

Macroblocks are the basic unit of coding for motion compensation. Motion vectors are determined for the luminance Macroblock and the motion vectors for chrominance are determined from the luminance motion vectors.

In Frame DCT coding, each Block is composed of lines from both fields (odd and even lines) in the same order they are displayed (i.e., odd, even, odd, . . .). The Macroblock is divided into four Blocks as always. In Field DCT coding, each Block is composed of lines from only odd or only even lines.

Macroblocks do not have a header. Their encoding proceeds sequentially within a Slice. From Figure 8.3 it is clear that the addresses of Macroblocks are not necessarily sequential. If the address difference is greater than 33, the 11-bit binary string '0000 0001 000', **macroblock_escape,** indicates that 33 should be added to the value indicated by **macroblock_address_increment,** a 1- to 11-bit variable length code. The increment value ranges from 1 to 33.

A 5-bit **quantizer_scale_code** follows next in the Macroblock. The presence of this parameter in each Macroblock is important—it allows quantization to be changed in every Macroblock. This means that an MPEG 2 Video Intra frame is much more efficient than an original JPEG-coded image in

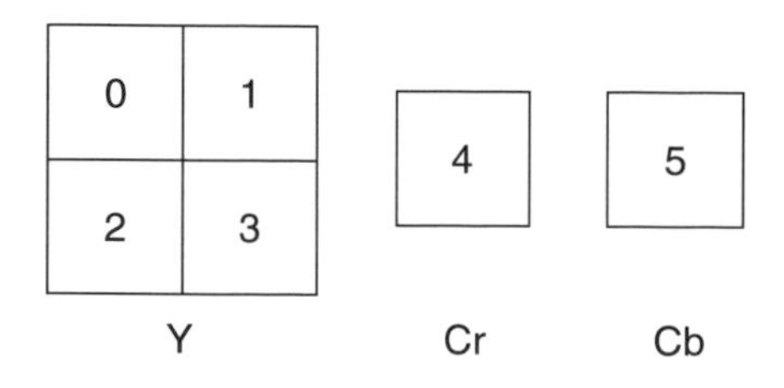

Figure 8.4 Macroblock Structure for 4:2:0

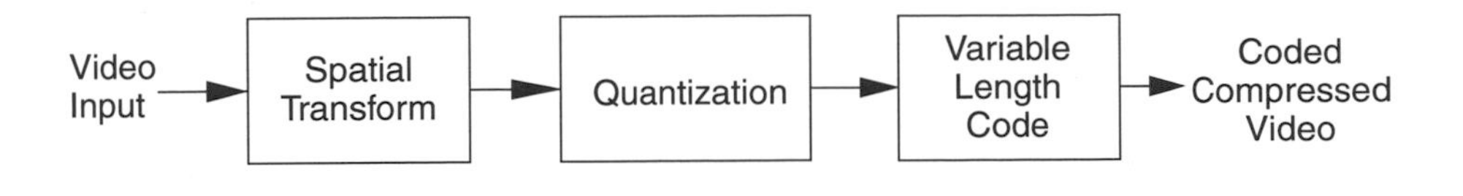

Figure 8.5 Spatial Compression Steps

which the quantization had to be set once for the picture (JPEG corrected this in later versions). Finally, there is a **marker_bit,** always '1'.

Block

Blocks are 8-by-8 pixels and are the smallest syntactic unit of MPEG Video. Blocks are the basic unit for DCT coding.

8.5 Detailed MPEG 2 Coding of Pictures

As previously noted, MPEG 2 Video compression consists of a spatial compression step and a temporal compression step. These are discussed in the following sections.

> **Spatial Compression:** Figure 8.5 shows the three major steps in spatial compression. First, the image pixels are transformed by an 8-by-8 block Discrete Cosine Transform (DCT). Next, the transform coefficients are quantized. Finally, the quantized coefficients are encoded with a variable length code (VLC). Note that the transform step and the variable length coding step are one-to-one mappings; therefore, the only losses in the spatial compression are generated in the quantization step.

8.5.1 I Pictures

Spatial Transform

Before starting a detailed discussion of the DCT, it is important to understand the role of the transform. The input to the transform is an 8-by-8 array of pixels from an image in which the intensity value of each pixel in each color plane ranges from 0 to 255. The output is another array of 8-by-8 numbers. The spatial transform takes the 8-by-8 element image Block and transforms it into the 8-by-8 transform coefficient Block, which can be coded with vastly fewer bits than can the original input Block.

The (0,0) transform coefficient is special. It represents the average value of the 64 input pixels and is called the DC value. As one moves to the right in the horizontal direction, or down in the vertical direction, the transform characterizations are said to be of increasing spatial frequency, where *spatial frequency* is a measure of the edge content or how close things are together. High spatial frequency corresponds to edges or where things are closer together. As with other similar one-dimensional counterparts, there is an inverse relationship between things being close together and high spatial frequencies. The DCT is effective because it tends to concentrate the energy in the transform coefficients in the upper-left corner of the array, which are the lower spatial frequencies.

The DCT: The N-by-N block DCT is defined as follows:

$$F(u,v) = \frac{2}{N} * C(u) * C(v) * \sum_{x=0}^{N-1} \sum_{y=0}^{N-1} f(x,y) * \cos\left[\frac{(2*x+1)*u*\pi}{2*N}\right] * \cos\left[\frac{(2*y+1)*v*\pi}{2*N}\right]$$

with u, v, x, y equal to $0, 1, \ldots,$ and 7, where x and y are spatial coordinates in the sample domain, u and v are coordinates in the transform domain, and

$$C(u), C(v) = \begin{cases} 1/\sqrt{2} & \text{for} \ldots u,v = 0 \\ 1 & \text{otherwise} \end{cases}$$

This can be particularized to the 8-by-8 case by setting $N = 8$.

Final Note about DCT: The optimum transform would be a Karhunen Loeve (KL) transform; however, the KL transform is very computationally intensive and data dependent. It turns out that if the underlying process to be transformed is first-order Markov, the DCT is nearly equivalent to the KL with much less computation required. Video is not quite first-order Markov, but sufficiently close so that, typically, DCT works very well. This is the reason the DCT is used in all major Video compression standards today.

Quantization

The second step of spatial compression is *quantization* of the transform coefficients, which reduces the number of bits used to represent a DCT coefficient. Because of the multiples used in the DCT, even though the input values are 8 bits, the DCT coefficients could be 11 bits if they were represented exactly. However, this would be extremely wasteful. Quantization is implemented by dividing the transform coefficient by an integer and then rounding to the nearest integer.

The integer divisor of each DCT coefficient is comprised of two parts. The first part is unique for each coefficient in the 8-by-8 DCT matrix. The set of these unique integers is itself a matrix and is called the Quant matrix.

The second part is an integer (**quantizer_scale**) that is fixed for at least each Macroblock. Thus, if Dct[i][j] is the DCT matrix, then the quantized DCT matrix, Qdct[i][j], is

$$Qdct[i][j] = 8 * Dct[i][j]/((\textbf{quantizer\_scale}) * Q[i][j]).$$

For the nonscalable parts of MPEG 2, which are of primary interest for current DBS, there are two default Quant matrices, one for Intra frames and a second for non-Intra frames. These two default matrices are shown in Tables 8.3 and 8.4.

Table 8.3 Quantization Matrix for Intra Blocks

	0	**1**	**2**	**3**	**4**	**5**	**6**	**7**
0	8	16	19	22	26	27	29	34
1	16	16	22	24	27	29	34	37
2	19	22	26	27	29	34	34	38
3	22	22	26	27	29	34	37	40
4	22	26	27	29	32	35	40	48
5	26	27	29	32	35	40	48	58
6	26	27	29	34	38	46	56	69
7	27	29	35	38	46	56	69	83

Table 8.4 Quantization Matrix for Non-Intra Blocks

	0	**1**	**2**	**3**	**4**	**5**	**6**	**7**
0	16	16	16	16	16	16	16	16
1	16	16	16	16	16	16	16	16
2	16	16	16	16	16	16	16	16
3	16	16	16	16	16	16	16	16
4	16	16	16	16	16	16	16	16
5	16	16	16	16	16	16	16	16
6	16	16	16	16	16	16	16	16
7	16	16	16	16	16	16	16	16

Variable Length Code

Zig Zag Scan: The next step in spatial compression is to map the quantized DCT coefficients into a one-dimensional vector that will be the stub for the Variable Length tables. This mapping is called a *zig zag scan.* Two zig zag scans are shown in Figures 8.6 and 8.7, one for progressive and the other for interlaced.

Codebook

After the quantized DCT coefficients have been mapped onto a one-dimensional vector, this vector becomes the stub for a variable length codebook. The purpose of the VLC is to minimize the average number of bits required to code this vector. Those readers who have studied Morse code have already encountered variable length code. The Morse code is based on the frequency of the occurrence of letters in the English language. For example, the letter "e" occurs about 13% of the time, is the most frequent, and thus has the code dot. The VLC for representing the zig zag output is also based on frequency of occurrence.

The VLCs used in MPEG 2 Video are known in Information Theory as Huffman codes. Huffman codes are compact codes, which means that if the

V \ U	0	1	2	3	4	5	6	7
0	DC 0	1	5	6	14	15	27	28
1	2	4	7	13	16	26	28	42
2	3	8	12	17	26	30	41	43
3	9	11	18	24	31	40	44	53
4	10	19	23	32	38	45	52	54
5	20	22	33	38	46	51	55	60
6	21	34	37	47	50	56	59	61
7	35	36	48	49	57	58	62	63

Figure 8.6 Zig Zag Scan for Progressive Frames

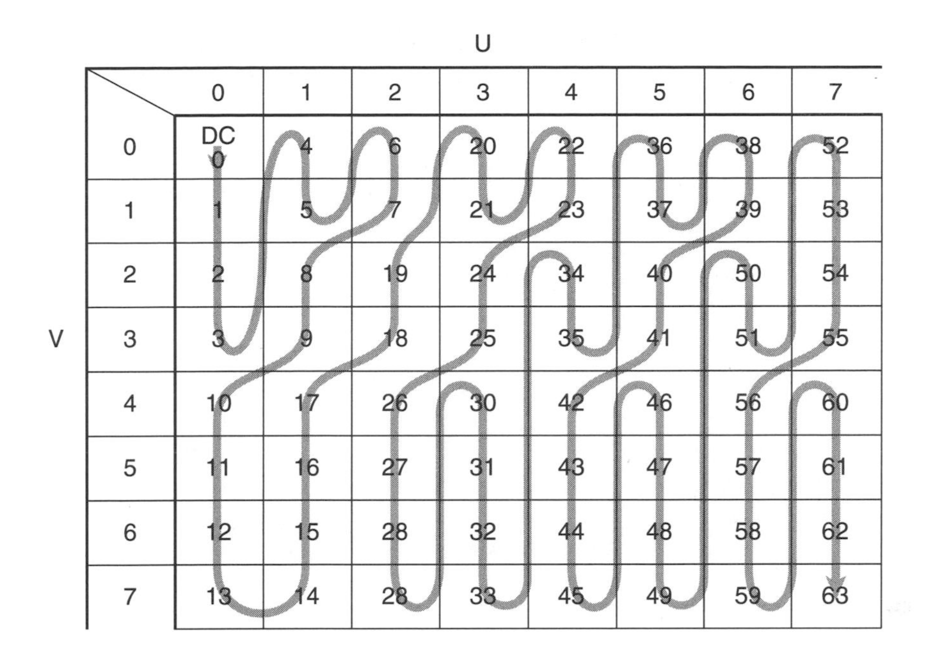

Figure 8.7 Zig Zag Scan for Interlaced Frames

underlying statistics are constant, no other code can have a shorter average length.

The Huffman codebook for DCT coefficients is based on the concept of run-length coding. In this technique, the number of consecutive zeroes becomes part of the codebook stub. It is unusual in that the stub includes not only a run length, but also the amplitude of the coefficient that ended the run. As an example, consider Table 8.5, which contains the first few entries from Table B-14 of ISO 13818-2.

Study of Table 8.5 shows the importance of getting the first few entries of the codebook correct. By just the tenth code, the code is seven bits long, compared with two bits for the shortest code—an increase of 3.5. By the end of Table B.14, the entries are 17 bits long!

8.5.2 P Pictures

P Pictures are predicted from prior I or P frames. There may or may not be intervening B Pictures. The prior I or P frame used to predict a P Picture is called the *reference picture*. The prediction of a P Picture is comprised of two separate steps: Motion Compensation and Residual Image coding.

Table 8.5 The Shortest Ten of DCT Huffman Codes

Variable Length Code[1]	Run (of zeroes)	Level
10	end of block	
1s[2]	0	1
11s	0	1
011s	1	1
0100s	0	2
0101s	2	1
0010 1s	0	3
0011 1s	3	1
0011 0s	4	1
0001 10s	1	2

[1]The "s" concatenated at the end of each code indicates the sign bit of the level, "0" for positive and "1" for negative.

[2]This code is used for the DC coefficient.

Motion Compensation

If the video from a camera pointed at the sky was being encoded, there would be very little change in the pixel values from frame to frame. If a 16-by-16 pixel block were being encoded, its values could be predicted almost exactly from the same area of the reference frame. Thus, we do not have to code the block at all. Instead, we could send a code that tells the decoder to use only the same values from the prior frame. Figure 8.8 depicts motion compensation.

The Macroblock to be predicted has a specific set of coordinates and there is a Macroblock in the input picture that corresponds to this location. Most practical encoders (the encoding process is not defined in ISO 133818-2) use the Macroblock in the input picture at that location to search the reference image in that location for the best fit.

The coordinate offsets in x and y are called motion vectors. Thus, if the motion vectors are coded with a VLC and sent to the decoder, the decoder will locate the coordinate of the current Macroblock being decoded and offset it by the motion vector to select a prediction Macroblock from the reference picture. The 16-by-16 Macroblock of pixels in the reference frame at this offset location becomes the initial prediction for this Macroblock (first step).

It is important that the prediction is where, in the reference frame, the Macroblock came from, rather than where a Macroblock is going. The latter

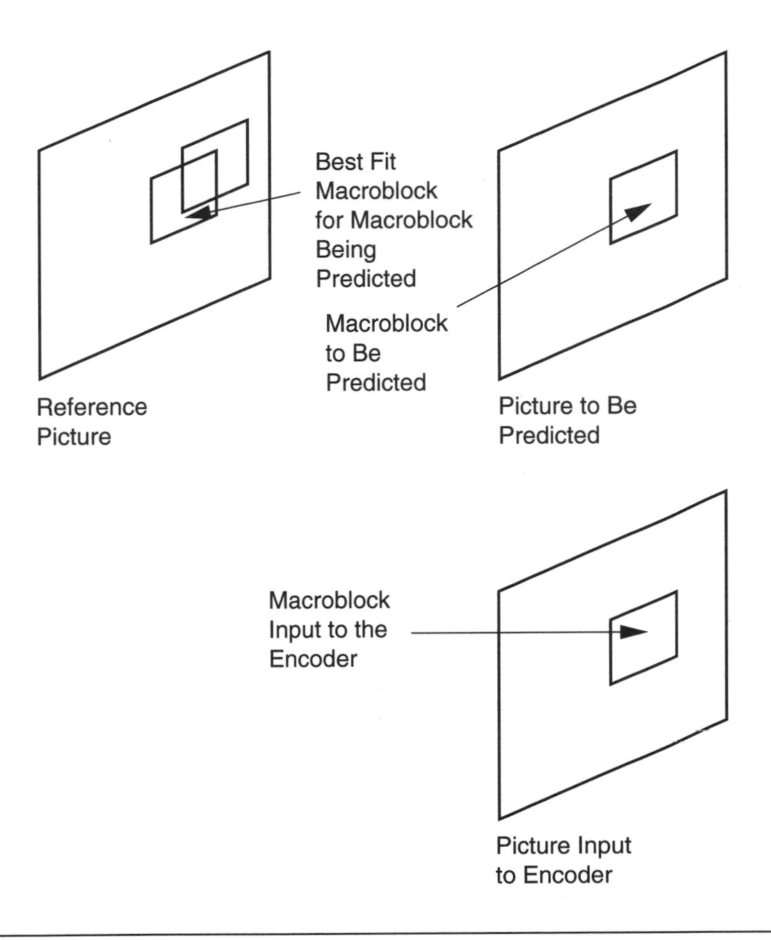

Figure 8.8 Motion Compensation for P Pictures

viewpoint would lead to overlaps and cracks in the predicted picture. Three predictions are made: forward prediction, backward prediction, and an average of the forward and backward prediction. The encoder can select which of these is the most efficient.

Residual Image Coding

The quality of the prediction can vary widely. As one example, if a scene was cut between the picture being predicted and the reference picture, there is really no match with the reference frame and the Macroblock has to be coded Intra even though it is in a P Picture. In other cases, the match may be reasonably good overall but poor in a particular location of the Macroblock.

In this case, the reference Macroblock is subtracted from the input Macroblock to form a residual image. This residual image is then coded like the spatial compression shown in Figure 8.5 and discussed previously under I Pictures. In the decoder, the prediction for the Macroblock from motion compensation is first made and then the decoded residual image is added to create the final prediction.

8.5.3 B Pictures

B Pictures are bidirectionally interpolated from reference pictures that precede and succeed the display of the frame being predicted.

Motion Compensation

The use of B Pictures is somewhat controversial, since those seeking maximum compression need them, while those editing these compressed materials have problems because access to frames is required after the frame to be predicted.

Maximum compression is achieved using B Pictures because the residual images of the B Pictures can be heavily quantized without seriously degrading the image quality. This is possible because several pictures of relatively poor quality can be put between "bookend" pictures of good quality without offending the HVS.

Figure 8.9 shows the bidirectional prediction of a B Picture Macroblock.

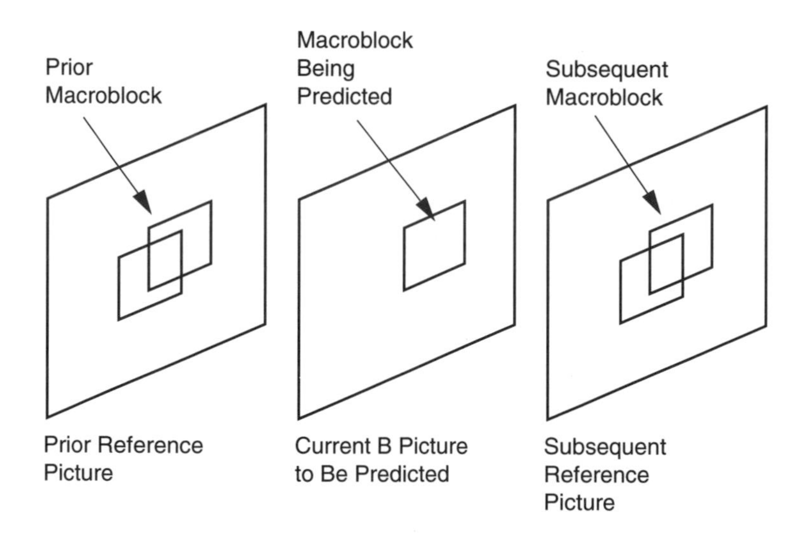

Figure 8.9 B Picture Motion Prediction

Residual Image Coding

As with P Pictures, a residual image is formed for each Macroblock. These are heavily quantized because a high compression ratio is required for B Pictures.

8.5.4 Coded Block Pattern

The Coded Block Pattern (CBP) is a VLC of 3 to 9 bits that represents a fixed-length, 6-bit code that tells the decoder which Blocks of the Macroblock have nonzero coefficient information. Each Block of a Macroblock is assigned a value; if all the coefficients of a Block are 0, the value is 0. Otherwise, the value is assigned according to this list.

Block	**Coded_block_pattern** Value
0	32
1	16
2	8
3	4
4	2
5	1

The nonzero values are added to form the CBP and then coded by a VLC. Note that if all Blocks have nonzero values, CBP = 63. If only the luminance Blocks have nonzero coefficients, CBP = 60. It is interesting to note that the value 60 has the shortest VLC (see Table B.9 of ISO 13818-2), which means it is the most frequently occurring value.

8.6 The Video Decoding Process

This section uses the preceding information to describe a complete decoding process.

8.6.1 Recovering the 8-by-8 Pixel Blocks

The decoding process always reverses the encoding process. Thus, the first step is to parse out each of the entities represented by a VLC. The VLC is then inverted to recover the 64-element vector that was coded by the VLC, $QFS[n]$. An inverse scan then maps $QFS[n]$ into the 8-by-8 array, $QF[v][u]$. Inverse quantization creates the array $F[v][u]$. Finally, the inverse DCT of $F[v][u]$ forms the 8-by-8 output pixel array, $f[x][y]$.

8.6.2 Variable Length Decoding

Variable length decoding is the inverse of the Variable Length Coding process. Since VLC is a 1:1 mapping, the inverse VLC is unique.

DC Component of Intra Blocks

The DC component of the DCT for Intra Blocks is treated differently from all other DCT components. The technique is a form of differential Pulse Code Modulation (PCM). At the beginning of each Slice of an Intra picture, a predictor is set for the DC value for each of the color planes.

In the encoder, a VLC is used to code the parameter **dct_dc_size.** If **dct_dc_size** is 0, the DC coefficient for that block is the predictor. If **dct_dc_size** is not 0, its value denotes the bit length of a following fixed-length code (up to 11-bits long). It is a differential value that is added to the predictor to create the DC coefficient. The new predictor is this DC value. The reset value of the DC predictor is derived from the parameter **intra_dc_precision.**

The predictors are defined as **dc_dct_pred[cc].** There is a 2-bit parameter called **cc** (for color component) that takes on the value '00' for luminance blocks, '01' for Cb, and '10' for Cr. Separate codebooks are utilized for luminance **dct_dc_size** Blocks and chrominance **dct_dc_size** Blocks.

Other DCT Coefficients

The non-Intra coefficients are decoded from other VLC codebooks. The following are three possibilities.

1. *End of block*—In this case, there are no more nonzero coefficients in the block and all the remaining coefficients are set to 0.
2. *A normal coefficient*—This is a combined value of run length and level followed by a single bit designating the sign of the level.
3. *An escape code*—In this case, special provisions are employed to code the run and level.

8.6.3 Inverse Scan

The decoded VLC codes represent a one-dimensional version of the quantized DCT coefficients. In order to inverse quantize these values and then invert the DCT, the one-dimensional vector must be mapped into a two-dimensional matrix. The process of doing this mapping is called the *inverse scan.*

Table 8.6 shows this mapping for **alternate_scan** = 0. Note that each entry in the table indicates the position in the two-dimensional array where the one-dimensional index is located. If one examines Figure 8.6, it can be seen

Table 8.6 Inverse Scan for **alternate_scan** = 0

	0	**1**	**2**	**3**	**4**	**5**	**6**	**7**
0	0	1	5	6	14	15	27	28
1	2	4	7	13	16	26	29	42
2	3	8	12	17	25	30	41	43
3	9	11	18	24	31	40	44	53
4	10	19	23	32	39	45	52	54
5	20	22	33	38	46	51	55	60
6	21	4	37	47	50	56	59	61
7	35	36	48	49	57	58	62	63

that it is identical to Table 8.6; however, in one case the conversion is being made from two dimensions to one dimension, while in the other the reverse is truc.

Similarly, the inverse for **alternate_scan** = 1 is the same as Figure 8.7. When quantization matrices are downloaded, they are encoded in the bitstream in this same scan order.

8.6.4 Inverse Quantization

The two-dimensional array of coefficients $QF[v][u]$ is Inverse Quantized to produce the reconstructed DCT coefficients. The process is essentially multiplication by the quantizer step size. This step size is modified by two different entities: (1) a weighting matrix ($W[w][v][u]$) modifies the step size within a Block, usually affecting each spatial frequency position differently, and (2) a scalar value (**quant_scale**) so that the step size can be modified at the cost of a few bits. The output of the first step of the Inverse Quantization is $F''[u][v]$.

Intra-DC Coefficients

The DC coefficients for Intra-coded Blocks are coded differently from all other coefficients. Thus, for Intra Blocks, $F''[0][0]$ = **intra_dc_mult** $*$ $QF[0][0]$.

Other Coefficients

All other Intra coefficients and all non-Intra coefficients are coded as described in the following sections.

Weighting Matrices

For MP@ML, two matrices are used, one for Intra and one for non-Intra. W[w][v][u] represents the weighting matrices (see Tables 8.3 and 8.4). The parameter w is determined from the parameter **cc**. For 4:2:0, **cc** = 0 for Intra Blocks and **cc** = 1 for non-Intra Blocks. For 4:2:2 and 4:4:4 for luminance, $w = 0$ for Intra Blocks if **cc** = 0, for luminance and non-Intra Blocks, if **cc** = 0, $w = 1$; for chrominance and Intra Blocks, if **cc** = 1, $w = 2$, and for chrominance and non-Intra Blocks, if **cc** = 1, $w = 3$.

Quantizer Scale Factor

The parameter **quant_scale** is needed in the inverse quantization process. The 1-bit flag, **q_scale_type,** which is coded in the picture extension, and the parameter **quantizer_scale_code** (contained in the Slice header) determine **quant_scale.** If **q_scale_type** = 0, then **quant_scale** equals two times the **quantizer_scale_code.** If **q_scale_type** = 1, **quant_scale** equals **quantizer_scale_code** for **quantizer_scale_code** from 1 to 8. For **quantizer_scale_code** from 9 to 16, **quant_scale** = 2 * **quantizer_scale_code** – 8. For **quantizer_scale_code** from 17 to 24, **quant_scale** = 4 * **quantizer_scale_code** – 40. For **quantizer_scale_code** from 25 to 31, **quant_scale** = 8 * **quantizer_scale_code** – 136.

Reconstruction of the DCT Coefficient Matrix

The first step in this reconstruction is to form $F''[v][u]$, where

$$F''[v][u] = (2 * QF[v][u] + k) * (k * W[w][v][u]) * (\mathbf{quantizer\_scale}))/32,$$

where $k = 0$ for Intra Blocks and $k = \text{sine}[QF[v][u]]$ for non-Intra Blocks.

Saturation

For the inverse DCT to work properly, the input range must be restricted to the range $2{,}047 \geq F'[v][u] \geq -2{,}048$. Thus,

$$F'[u][v] = \begin{cases} 2047, & F'' > 2047 \\ F''[u][v], & 2047 \geq F''[u][v] \geq -2048 \\ -2048, & F''[u][v] \leq -2048 \end{cases}$$

This function, which electronics engineers would call a *limiter,* is called saturation in MPEG 2.

Mismatch Control

The sum of the coefficients to the Inverse DCT must be odd. Thus, all the coefficients of $F'[v][u]$ are summed. If the sum is odd, all the coefficients of

$F[v][u]$ are the same as $F'[v][u]$. If the sum is even, $F[7][7]$ is modified. If $F'[7][7]$ is odd, $F[7][7] = F'[7][7] - 1$; if $F'[7][7]$ is even, $F[7][7] = F'[7][7] + 1$. This ensures that the sum of the coefficients input to the inverse DCT is odd.

8.6.5 Inverse DCT

Finally, the inverse DCT recreates the spatial values for each Block using the following equation:

$$f(x,y) = 2/N \sum_{u=0}^{N-1} \sum_{v=0}^{N-1} C(u)C(v)F(u,v)\cos[(2x+1)u\pi/2N]\cos[(2y+1)v\pi/2N]$$

where N is set to 8 for the 8-by-8 blocks.

8.6.6 Motion Compensation

Figure 8.10 is an overview of the motion compensation output. The output of the Inverse DCT, $f[y][x]$, is added to the motion prediction output $p[y][x]$. The sum is then limited to the range 0 to 255 by the Saturation step. The output of the Saturation step is the decoded pixels for display.

Several things should be noted about Figure 8.10. First, even though the predictions are for Macroblocks, $p[y][x]$ is the appropriate Block of the predicted Macroblock. Second, for the chrominance axes, the decoded pixels out are still 8-by-8 pixels. The decoder must interpolate them to 16-by-16 pixels prior to display. Finally, it should be noted that the output of the Inverse DCT, $f[y][x]$, is from the result of the coding of the residual image; this is why it is added to the prediction.

Combinations of Predictions

The prediction Block, $p[y][x]$ in Figure 8.10, can be generated from as many as four different types, which are discussed in the following sections.

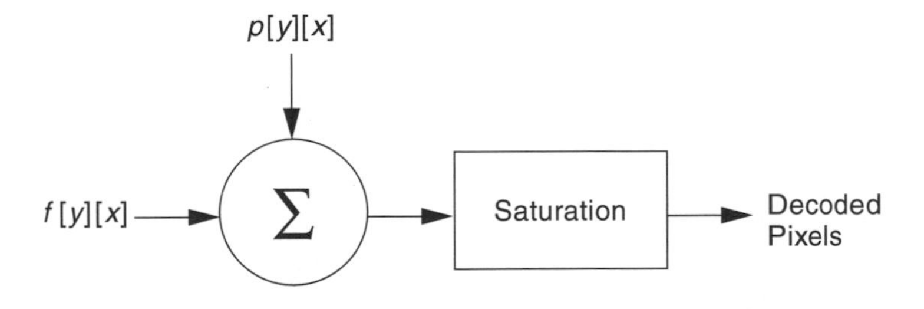

Figure 8.10 Overview of Motion Compensation Output

Simple Frame Predictions

The transform data $f[y][x]$ is either field organized or frame organized depending on **dct_type.** If it specifies frame prediction, then for simple frame prediction the only function the decoder has to perform is to average the forward and backward predictions in B Pictures.

If **pel_pred_forward[y][x]** is the forward prediction and **pel_pred_backward[y][x]** is the corresponding backward prediction, then the final prediction is:

pel_pred[y][x] = (pel_pred_forward[y][x] + pel_pred_backward[y][x])/2

For MP@ML, all the predictors are 8-by-8 pixels.

Simple Field Predictions

In the case of simple field predictions, the only processing required is to average the forward and backward predictions as for the simple frame predictions just described.

16-by-8 Motion Compensation

In this prediction mode, separate predictions are formed for the upper 16-by-8 region of the Macroblock and the lower 16-by-8 region of the Macroblock. The predictions for the chrominance components are 8 samples by 4 lines for MP@ML.

Dual Prime

In Dual Prime mode, two predictions are formed for each field in a manner similar to that for backward and forward prediction of B Pictures. If **pel_pred_same_parity** is the prediction sample from the same parity field, and **pel_pred_opposite_parity** is the corresponding sample from the opposite parity, then the final prediction sample is formed as:

pel_pred[y][x] = (pel_pred_same_parity[y][x] + pel_pred_opposite_parity[y][x])/2

In the case of Dual Prime prediction in a frame picture, the predictions for the chrominance components of each field for MP@ML are 8 samples by 4 lines. In the case of Dual Prime prediction in a field picture, the predictions for chrominance components are 8 samples by 8 lines for MP@ML.

8.6.7 Skipped Macroblocks

In Skipped Macroblocks (where **macroblock_address_increment** is greater than 1), the decoder has neither DCT coefficients nor motion vector information. The decoder has to form a prediction for these Macroblocks that is used as final decoded sample values. The handling of Skipped Macroblocks is different for P Pictures and B Pictures and for field pictures and frame pictures. Macroblocks for I Pictures cannot be skipped.

P Field Pictures

The prediction is made as if **field_motion_type** is "Field based" and the prediction is made from the field of the same polarity as the field being predicted. Motion vector predictors and motion vectors are set to 0.

P Frame Pictures

The prediction is made as if **frame_motion_type** is "Frame based." Motion vector predictors and motion vectors are set to 0.

B Field Pictures

The prediction is made as if **field_motion_type** is "Field based" and the prediction is made from the field of the same polarity as the field being predicted. The direction of the prediction (forward/backward/bidirectional) is the same as the previous Macroblock. The motion vector predictors are unaffected. The motion vectors are taken from the appropriate motion vector predictors, dividing both the horizontal and vertical components by two scales of the chrominance motion vectors.

B Frame Pictures

The prediction is made as if **frame_motion_type** is "Frame based." The direction of the prediction (forward/backward/bidirectional) is the same as the previous Macroblock. The motion vector predictors are unaffected. The motion vectors are taken from the appropriate motion vector predictors, dividing both the horizontal and vertical components by two scales of the chrominance motion vectors.

8.7 Prediction Modes

There are two major prediction modes: field prediction and frame prediction. (For more detailed information, see Sections 7.1 to 7.5 of ISO 13818-2.)

In field prediction, predictions are made independently for each field by using data from one or more previously decoded fields. Frame prediction forms a prediction for the frame from one or more previously decoded frames. The fields and frames from which predictions are made may themselves have been decoded as either field or frame pictures.

Within a field picture, all predictions are field predictions. However, in a frame picture either field or frame predictions may be used (selected on a Macroblock-by-Macroblock basis).

8.7.1 Special Prediction Modes

In addition to the two major classifications, there are two special prediction modes: 16-by-8 and Dual Prime. Because Dual Prime is used only when there are no B Pictures, it will not be discussed further here.

In 16-by-8 motion compensation, two motion vectors are used for each Macroblock. The first motion vector is used for the upper 16-by-8 region, the second for the lower 16-by-8 region. In the case of bidirectionally interpolated Macroblocks, a total of four motion vectors are required because there will be two for the forward prediction and two for the backward prediction. A 16-by-8 prediction is used only with field pictures.

8.7.2 Field and Frame Prediction

The prediction for P and B frames is somewhat different for fields and frames.

Field Prediction

In P Pictures, prediction is made from the two most recently decoded reference fields. In the simplest case, the two reference fields are used to reconstruct the frame, regardless of any intervening B frames.

The reference fields themselves may have been reconstructed by decoding two field pictures or a single frame picture. When predicting a field picture, the field being predicted may be either the top field or the bottom field.

Predicting the second field picture of coded frame uses the two most recently decoded reference fields and the most recent reference field is obtained from decoding the first field picture of the coded frame. When the field to be predicted is the bottom frame, this is the top field; when the field to be predicted is the top field, the bottom field of that frame is used. Once again, intervening B fields are ignored.

Field prediction in B Pictures is made from the two fields of the two most recently reconstructed reference frames, one before and one after the frame being predicted. The reference frames themselves may have been reconstructed from two field pictures or a single frame picture.

Frame Prediction

In P Pictures, prediction is made from the most recently reconstructed reference frame. Similarly, frame prediction in B Pictures is made from the two most recently reconstructed reference frames, one before and one after the frame being predicted.

8.7.3 Motion Vectors

Motion vectors are coded differentially with respect to previously decoded motion vectors. To decode the motion vectors, the decoder must maintain four motion vector predictors (each with a horizontal and vertical component) denoted *PMV* [*r*][*s*][*t*]. For each prediction, a motion vector *vector'* [*r*][*s*][*t*] is derived first. This is then scaled according to the sampling structure to give another motion vector, *vector*[*r*][*s*][*t*], for each color component.

The [*r*][*s*][*t*] Triplet

The parameter *r* determines whether the first motion vector in a Macroblock is indicated with a '0', the second, with a '1'. The parameter *s* indicates whether the motion vector is forward, with a '0', or backward, with a '1'. The parameter *t* indicates whether the motion vector represents a horizontal component, with a '0', or a vertical component, with a '1'.

Two types of information are decoded from the bitstream in order to make the prediction: (1) information on the delta to be added to the prediction and (2) which motion vector predictor to use.

Updating Motion Vector Predictors

Once the motion vectors for a Macroblock have been determined, it is necessary to update the motion vector predictors. This is necessary because these predictors may be used on a subsequent Macroblock.

If the **frame_motion_type** is:

Frame-based and the **macroblock_intra** is '1', then PMV[1][0][1] = PMV[0][0][1] and PMV[1][0][0] = PMV[0][0][0].

Frame-based and the **macroblock_intra** is '0', and **macroblock_motion_forward** and **macroblock_motion_backward** are both '1', then PMV[1][0][1] = PMV[0][0][1], PMV[1][0][0] = PMV[0][0][0], PMV[1][1][1] = PMV[0][1][1], and PMV[1][1][0] = PMV[0][1][0].

Frame-based and the **macroblock_intra** is '0', the **macroblock_motion_forward** = '1', and the **macroblock_motion_backward** = '0', then PMV[1][0][1] = PMV[0][0][1] and PMV[1][0][0] = PMV[0][0][0].

Frame-based and the **macroblock_intra** is '0', the **macroblock_motion_forward** = '0', and the **macroblock_motion_backward** = '1', then PMV[1][1][1] = PMV[0][1][1] and PMV[1][1][0] = PMV[0][1][0].

Frame-based and the **macroblock_intra** is '0', the **macroblock_motion_forward** = '0', and **macroblock_motion_backward** = '0', then PMV [*r*][*s*][*t*] = 0 for all *r*,*s*,*t*.

Field-based only, there are no predictors to update.

Dual Prime and **macroblock_intra** is '0', the **macroblock_ motion_ forward** = '1', and the **macroblock_motion_backward** = '0', then PMV[1][0][1] = PMV[0][0][1] and PMV[1][0][0] = PMV[0][0][0].

If the **field_motion_type** is:

Field-based and the **macroblock_intra** is '1', then PMV[1][0][1] = PMV[0][0][1] and PMV[1][0][0] = PMV[0][0][0].

Field-based and the **macroblock_intra** is '0', and **macroblock_ motion_forward** and **macroblock_motion_backward** are both '1', then PMV[1][0][1] = PMV[0][0][1], PMV[1][0][0] = PMV[0][0][0], PMV[1][1][1] = PMV[0][1][1], and PMV[1][1][0] = PMV[0][1][0].

Field-based and the **macroblock_intra** is '0', the **macroblock_ motion_forward** = '1', and the **macroblock_motion_backward** = '0', then PMV[1][0][1] = PMV[0][0][1] and PMV[1][0][0] = PMV[0][0][0].

Field-based and the **macroblock_intra** is '0', the **macroblock_ motion_forward** = '0', and the **macroblock_motion_backward** = '1', then PMV[1][1][1] = PMV[0][1][1] and PMV[1][1][0] = PMV[0][1][0].

Field-based and the **macroblock_intra** is '0', the **macroblock_ motion_forward** = '0', and **macroblock_motion_backward** = '0'.

Field-based and the **macroblock_intra** is '0', the **macroblock_ motion_forward** = '0', and **macroblock_motion_backward** = '0', then PMV[*r*][*s*][*t*] = 0 for all *r,s,t*.

For a 16-by-8 MC, there are no predictors to update.

Dual Prime and **macroblock_intra** is '0', the **macroblock_ motion_ forward** = '1', and the **macroblock_motion_backward** = '0', then PMV[1][0][1] = PMV[0][0][1] and PMV[1][0][0] = PMV[0][0][0].

Resetting Motion Vector Predictors

All motion vector predictors are reset to 0 in the following situations:

- At the start of each slice
- Whenever an Intra Macroblock is decoded that has no concealment motion vectors
- In a P Picture when a non-Intra Macroblock is decoded in which **macroblock_motion_forward** is 0
- In a P Picture when a Macroblock is skipped

Motion Vectors for Chrominance Components

For MP@ML, both the horizontal and vertical components of the motion vector are scaled by dividing by two:

$$vector[r][s][0] = vector'[r][s][0]/2$$
$$vector[r][s][1] = vector'[r][s][1]/2$$

Forming Predictions

Predictions are formed by reading prediction samples from the reference fields or frames. A particular sample is predicted by reading the corresponding sample in the reference field or frame offset by the motion vector.

A positive value of the horizontal component of a motion vector indicates that the prediction is made from samples in the reference field or frame that lie to the right of the samples being predicted. A positive value of the vertical component of a motion vector indicates that the prediction is made from samples in the reference field or frame that lie below the samples being predicted.

All motion vectors are specified to an accuracy of one-half sample. The one-half samples are calculated by simple linear interpolation from the actual samples.

In the case of field-based predictions, it is necessary to determine which of the two available fields to use to form the prediction. If **motion_vertical_field_select** is 0, the prediction is taken from the top reference field; if it is '1', from the bottom reference field.

CHAPTER NINE

MPEG 1 Audio Compression

The compact disc (CD) made digital audio a consumer product almost overnight. In a matter of two decades, it became difficult to even find a 33-rpm vinyl phonograph record. Consumers no longer had to put up with the annoying aspects of analog media such as flutter, wow, hiss, and other artifacts. Instead, they could have approximately an hour of flawless audio play on a small, compact, robust medium, with a player costing under $200.

To achieve its audio quality, the CD contains samples of an analog signal at 44.1 Ksamples per second and 16 bits per sample, or 705.6 Kbps. Because two analog signals are needed to create stereo, the 705.6 Kbps is multiplied by two, or 1,411.2 Kbps. If this is compared to the required services per transponder given in Chapter 8, it is clear that, if CD-quality audio is desired, the audio would take up too large a fraction of the total bit rate if not compressed by a significant amount.

The Compression Problem: DBS audio has 24-millisecond input frames, and samples each of the analog inputs at 48 Ksamples per second and 16 bits per sample. The output is 192 Kbps for both audio outputs combined.

$$\begin{aligned}\text{Arriving bits} &= 48 * 10^3 \frac{\text{samples}}{\text{second}} * 16 \frac{\text{bits}}{\text{sample}} * 24 * 10^{-3} \text{ seconds} \\ &* 2 \text{ signals} = 36,864 \text{ bits} \\ \text{Leaving bits} &= 96 * 10^3 \text{ bits/second/channels} * 2 \text{ channels} \\ &* 24 * 10^{-3} \text{ seconds} = 4608 \text{ bits}\end{aligned}$$

Therefore, audio compression must reduce the number of bits by 36,864/4608 = 8; in other words, a compression ratio of 8:1. MPEG Audio provides this compression.

It should be noted that while the MPEG Audio standard was developed as part of an integrated audio/video standard, the audio part can be used alone. In fact, one of the more surprising results of DBS service has been the popularity of the audio service. DIRECTV™ provides 28 "commercial-free" audio channels with the menu indicating the music type. At our house we frequently leave the music on during the day, just for background.

9.1 MPEG Audio Compression Overview

MPEG 1 Audio (ISO 11172-3) is oriented toward high-quality stereo audio. At the present time, all known DBS systems use Layer II of MPEG 1 Audio, at 192 Kbps, for stereo delivery.

MPEG 1 Audio provides a very large number of options in both sampling frequency, output bit rate, and three "layers." The MPEG 1 Audio Layers are not layers in the sense that one builds on the other. Rather, they are layers in the sense that a higher-numbered layer decoder can decode a lower-numbered layer. In this book, the discussion is limited to the particular options used in DBS systems.

ISO 11172-3 defines only a decoder. However, including an encoder description makes it easier to understand.

9.1.1 Encoding

Prior to MPEG 1 Audio, audio compression usually consisted of removing statistical redundancies from an electronic analog of the acoustic waveforms. MPEG 1 Audio achieved further compression by also eliminating audio irrelevancies by using psychoacoustic phenomena such as spectral and temporal masking. One way to explain this is that if the signal at a particular frequency is sufficiently strong, weaker signals that are close to this frequency cannot be heard by the human auditory system and, therefore, can be neglected entirely.

In general, the MPEG 1 Audio encoder operates as follows. Input audio samples are fed into the encoder. For DBS, these samples are at a sample rate of 48 Ksamples per second (for each of the stereo pairs), so this rate will be used exclusively in the rest of this chapter. Each sample has 16 bits of dynamic range. A mapping creates a filtered and subsampled representation of the input audio stream. In Layer II, these 32 mapped samples for each channel are called subband samples.

In parallel, a Psychoacoustic Model performs calculations that control quantizing and coding. Estimates of the masking threshold are used to perform this quantizer control. The bit allocation of the subbands is calculated on the basis of the signal-to-mask ratios of all the subbands. The maximum signal level and the minimum masking threshold are derived from a Fast Fourier Transform (FFT) of the sampled input signal.

Table 9.1 DBS Audio Parameters

Parameter	Value
Sampling rate	48 Ksamples per second for each of the stereo pairs
Transmit bit rate	192 Kbps
Stereo	Joint stereo/intensity stereo
Psychoacoustic Model	Psychoacoustic Model 1

Of the mono and stereo modes possible in the MPEG 1 Audio standard, Joint Stereo (left and right signals of a stereo pair coded within one bitstream with stereo irrelevancy and redundancy exploited) is used in Layer II—hence, DBS. Within Joint Stereo, the intensity stereo option is used. All DBS systems use 192 Kbps for this stereo pair.

9.1.2 Decoding

The basic structure of the audio decoder reverses the encoding process. Bitstream data is fed into the decoder. First, the bitstream is unpacked with the main data stream separated from the ancillary data. A decoding function does error detection if an **error_check** has been applied in the encoder. The bitstream data is then unpacked to recover the various pieces of information. A reconstruction function reconstructs the quantized version of the set of mapped samples. The inverse mapping transforms these mapped samples back into a Pulse Code Modulation (PCM) sequence. The output presentation is at 48 Ksamples per second for each of the two stereo outputs. The specific parameters utilized by DBS are shown in Table 9.1.

9.2 Description of the Coded Audio Bitstream

The basic building block of the MPEG 1 Audio standard is an Audio Frame. An *Audio Frame* represents a duration equal to the number of samples divided by the sample rate. For DBS, Layer II has 1,152 samples at a sample rate of 48 Ksamples per second. The frame period is then

$$\frac{1{,}152\ \dfrac{\text{samples}}{\text{frame}}}{48{,}000\ \dfrac{\text{samples}}{\text{second}}} = 24 \text{ milliseconds per frame}$$

Audio frames are concatenated to form audio sequences.

Audio Frame: An audio frame is a fixed-length packet in a packetized bitstream. In Layer II, it consists of 1,152 samples each of stereo audio input plus the rest of the bits comprising the packet. Each packet has a total of 4,608 bits. It starts with a syncword, and ends with the byte before the next syncword.

Each audio frame consists of four separate parts:

1. All packets in a packetized bitstream contain a **header** that provides synchronization and other information required for a decoder to decode the packet.
2. The **error_check** (optional) contains 16 bits that can detect errors in storage or transmission.
3. The **audio_data** is the payload of each packet. It contains the coded audio samples and the information necessary to decode them.
4. The **ancillary_data** has bits that may be added by the user.

Each of these is described in the following sections.

9.2.1 Header

The first 32 bits (4 bytes) of an Audio Frame are **header** information.

The **syncword** is the 12-bit string 0xFFF.

The **ID** is a 1-bit flag that indicates the identification of the algorithm; it is '1' for all DBS systems.

The layer parameter is the 2 bits, '10', that indicates Layer II.

The **protection_bit** is a 1-bit flag that indicates whether error detection bits have been added to the audio bitstream. It is a '1' if no redundancy has been added, '0' if redundancy has been added.

The **bitrate_index** is a 4-bit parameter that indicates the bit rate. In the case of DBS, the value is '1010' to indicate 192 Kbps for the stereo.

The **sampling_frequency** is a 2-bit parameter that indicates sampling frequency. For DBS, the value is '01', indicating 48 Ksamples per second.

The **padding_bit** is not required at 48 Ksamples per second.

The **private_bit** is for Private use.

The **mode** is a 2-bit parameter that indicates the channel mode. For DBS, the value is '01', indicating joint stereo. In Layer II, joint stereo is intensity stereo.

Figure 9.1 shows the operation of intensity stereo. For subbands less than **sb_bound** (bound), there are separate left and right audio bitstreams. For subbands of **sb_bound** and larger, a single bitstream that is the sum of the left and right values is transmitted. The left and right signal amplitude can be controlled by a different scalefactor.

> **bound** and **sblimit** are parameters used in **intensity_stereo** mode. In **intensity_stereo** mode **bound** is determined by **mode_extension.**
>
> The **mode_extension** is a 2-bit parameter that indicates which subbands are in **intensity_stereo.** All other subbands are coded in stereo. If the **mode_extension** is considered as a decimal number dec, the subbands dec * 4 + 4 to 31 are coded in intensity stereo. For example, if **mode_extension** is '11', dec = 3 and dec * 4 + 4 = 16. Thus, subbands 16 through 31 are coded in **intensity_stereo.**
>
> The **copyright** flag is 1 bit. A '1' means the compressed material is copyright protected.
>
> The **original/copy** is '0' if the bitstream is a copy, '1' if it is an original.
>
> The **emphasis** is a 2-bit parameter that indicates the type of deemphasis that is used. For DBS, the parameter is '01', indicating 50/15-microsecond deemphasis.

9.2.2 error_check

The **error_check** is an optional part of the bitstream that contains the Cyclical Redundancy Check (**crc_check**), a 16-bit parity-check word. It provides for error detection within the encoded bitstream (see Appendix E).

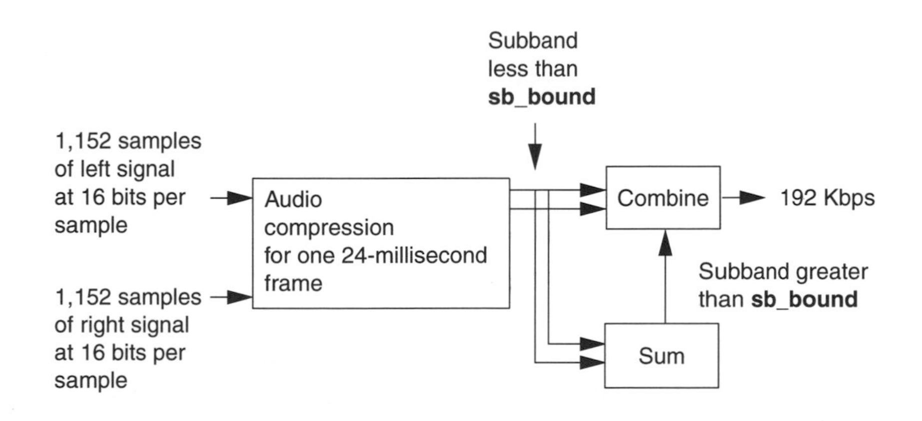

Figure 9.1 Intensity Stereo

9.2.3 audio_data, Layer II

The **audio_data** is the payload part of the bitstream. It contains the coded audio samples, and information on how these audio samples are to be decoded. The following are the key elements of **audio_data,** Layer II.

allocation[ch][sb]—[ch] indicates the left or right stereo channel, '0' for left and '1' for right. **[sb]** indicates the subband. **allocation[ch][sb]** is a 4-bit unsigned integer that serves as the index for the algorithm to calculate the number of possible quantizations (see Possible Quantization per Subband section later in this chapter).

scfsi[ch][sb]—This provides scalefactor selection information on the number of scalefactors used for **subband[sb]** in **channel[ch].** In general, each of the three 12 subband samples requires a separate scalefactor. In this case, **scfsc[sb]** = '00'. If all three of the parts can be represented by a single scalefactor, **scfsi[sb]** = '10'. If two scalefactors are required, **scfsi[sb]** = '01' if the first scalefactor is valid for parts 0 and 1 and the second for part 2; **scfsi[sb]** = '11' if the first scalefactor is valid for part 0 and the second for parts 1 and 2.

scalefactor[ch][sb][p]—[p] indicates one of the three groupings of subband samples within a subband. **scalefactor[ch][sb][p]** is a 6-bit unsigned integer which is the index to the scalefactor calculation in the Possible Quantization per Subband section later in this chapter.

grouping[ch][sb]—This is a 1-bit flag that indicates if three consecutive samples use one common codeword and not three separate codewords. It is true if the calculation in the Possible Quantization per Subband section creates a value of 3, 5, or 9; otherwise it is false. For subbands in **intensity_stereo** mode, the grouping is valid for both channels.

samplecode[ch][sb][gr]—[gr] is a granule that represents 3 * 32 subband samples in Layer II. **samplecode[ch][sb][gr]** is the code for three consecutive samples of **[ch]**, **[sb]**, and **[gr].**

sample[ch][sb][n]—This is the coded representation of the *n*th sample in **[sb]** of **[ch].** For the subbands here, **sb_bound,** and hence in **intensity_stereo** mode, this is valid for both channels.

9.2.4 ancillary_data

This MPEG 1 Audio bitstream has provisions for user supplied data. The number of ancillary bits (**no_of_ancillary_bits**) used must be subtracted from the total bits per frame. Since the frame length is fixed, this subtracts from the bits available for coding the audio samples and could impact audio quality.

9.3 Detailed Encoder

The MPEG 1 Audio algorithm is what is called a psychoacoustic algorithm. The following sections describe the four primary parts of the encoder.

9.3.1 The Filterbanks

The Filterbanks provide a sampling of the frequency spectrum for each of the 32 input samples. These Filterbanks are critically sampled, which means there are as many samples in the analyzed (frequency) domain as there are in the time domain. For the encoder, the Filterbank is called the "Analysis Filterbank" (see details in next section); in the decoder, the reconstruction filters are called the "Synthesis Filterbank" (see details at end of this chapter).

In Layer II, a Filterbank with 32 subbands is used. In each subband, there are 36 samples that are grouped into three groups of 12 samples each.

Input High-Pass Filter

The encoding algorithms provide a frequency response down to D.C. It is recommended that a high-pass filter be included at the input of the encoder with a cut-off frequency in the range of 2 Hz to 10 Hz. The application of such a filter avoids wasting precious coding bits on sounds that cannot be heard by the human auditory system.

Analysis Subband Filterbank

An analysis subband Filterbank is used to split the broadband signal with sampling frequency f_s into 32 equally spaced subbands with sampling frequencies $f_s/32$. The algorithm for this process with the appropriate formulas is given in the following algorithm description.

Analysis Filterbank Algorithm

1. Shift 32 new audio samples into a shift register.

512 Stage Shift Register Contains the **X Vector**

Audio Samples In →	0	1	• • •	511

Shift 32 Audio Samples in From Left. Shift Oldest 32 Audio Samples Out of the Right

2. Window by multiplying each of the **X** values by an appropriate coefficient.

$$Z_i = C_i * X_i \qquad i = 0, 1, \ldots 511 \tag{9.1}$$

3. Next form a 64-component vector **Y,** as follows:

$$
\begin{aligned}
Y_i &= \sum_{j=0}^{7} Z_{i+64*j} \\
Y_0 &= Z_0 + Z_{64} + \ldots + Z_{448} \\
Y_1 &= Z_1 + Z_{65} + \ldots + Z_{449} \\
&\vdots \\
Y_{63} &= Z_{63} + Z_{127} + \ldots + Z_{511}
\end{aligned}
\tag{9.2}
$$

4. Create subband outputs:

$$
\begin{bmatrix} S_0 \\ S_1 \\ \cdot \\ \cdot \\ \cdot \\ S_{31} \end{bmatrix} = \begin{bmatrix} M_{00} & M_{01} & \cdot & \cdot & \cdot & M_{0,63} \\ M_{10} & M_{11} & \cdot & \cdot & \cdot & M_{1,63} \\ \cdot & \cdot & \cdot & \cdot & \cdot & \cdot \\ \cdot & \cdot & \cdot & \cdot & \cdot & \cdot \\ \cdot & \cdot & \cdot & \cdot & \cdot & \cdot \\ M_{31,0} & M_{31,1} & \cdot & \cdot & \cdot & M_{31,63} \end{bmatrix} * \begin{bmatrix} Y_0 \\ Y_1 \\ \cdot \\ \cdot \\ \cdot \\ Y_{63} \end{bmatrix} \tag{9.3}
$$

$$
M_{ik} = \cos[\{(2*i+1)*(k-16)*\pi\}/64], \text{ for } i = 0 \text{ to } 31, \text{ and } k = 0 \text{ to } 63 \tag{9.4}
$$

Figure 9.2 shows the values D_i that are used in the Synthesis Subband Filter. The C_i values used in Equation (9.1) are obtained by dividing the *D* values by 32 for each index. Tables in the standard represent both the *C* and *D* values. I find the curves (see figure) more compact and instructive.

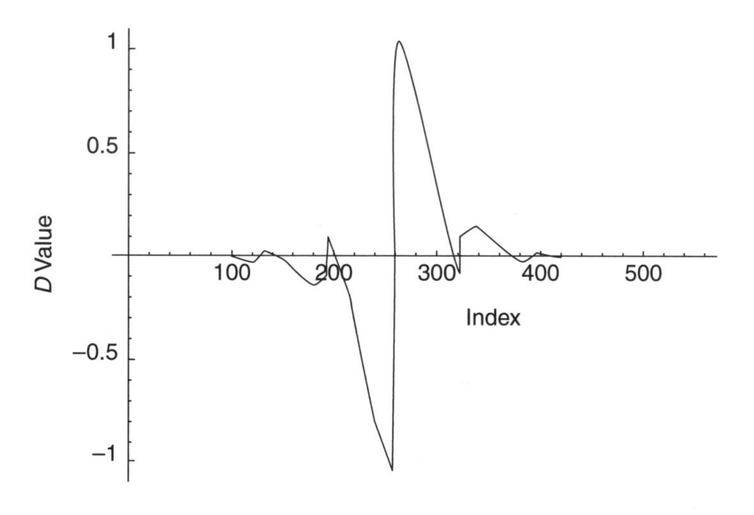

Figure 9.2 Parameter *D* Coefficients

9.3.2 Psychoacoustic Model

Psychoacoustic Model 1 is used with all known DBS systems and will be the only Psychoacoustic Model discussed. For Psychoacoustic Model 1, the FFT has a 1,024 sample window (see Analysis that follows). While this is less than the number of samples in a frame, it does not cause serious impact on quality.

The end result of the Psychoacoustic Model is the following signal-to-mask ratio:

$$SMR_{sb}(n) = L_{sb}(n) - LT_{\min}(n) \text{ dB} \tag{9.5}$$

which must be computed for every subband n. $L_{sb}(n)$ is the sound pressure level (SPL) in each subband n. $LT_{\min}(n)$ is the minimum masking level in each subband n. In the following sections, we show how each of these is computed.

FFT Analysis: The masking threshold is derived from the FFT output. A form of raised cosine filter, called a Hann window, filters the input PCM signal prior to calculating the FFT. The windowed samples and the Analysis Filterbank Outputs have to be properly aligned in time. A normalization to the reference level of 96 dB SPL has to be done in such a way that the maximum value corresponds to 96 dB.

Sound Pressure Level Calculation

Equation (9.5) requires calculation of the SPL $L_{sb}(n)$ in subband n. $L_{sb}(n)$ is determined from

$$L_{sb}(n) = MAX\left[X(k), \{20 * \log[scf_{\max}(n) * 32{,}768] - 10\}\right] \text{ dB} \tag{9.6}$$

$X(k)$ in subband n, where $X(k)$ is the sound pressure level with index k of the FFT with the maximum amplitude in the frequency range corresponding to subband n. The expression $scf_{\max}(n)$ is the maximum of the three scalefactors of subband n within an Audio Frame. The "–10 dB" term corrects for the difference between peak and RMS level. The SPL $L_{sb}(n)$ is computed for every subband n.

The second step in calculating Equation (9.5) is to calculate $LT_{\min}(n)$, which turns out to be a rather involved process. The minimum masking level, $LT_{\min}(n)$, in subband n is determined by the following expression:

$$LT_{\min}(n) = \underset{f(i) \text{ in subband } n}{Min}\left[LT_g(i)\right] \text{ dB} \tag{9.7}$$

Table 9.2 FFT Parameters

Layer II Parameter	Value
Transform length	1,024 samples
Window size if $f_s = 48$ kHz	21.3 ms
Frequency resolution	**sampling_frequency**/1,024
Hann window, $h(i)$:	$0 \leq i < N-1$
$h(i) = \sqrt{8/3} * 0.5 * \{1 - \cos[(2 * \pi * i)/N]\}$	
Power density spectrum $X(k)$:	
$X(k) = 10 * \log_{10}\left[\left\vert \frac{1}{N} \sum_{l=0}^{N-1} h(l) * s(l) * e^{(-j*k*l*2*\pi/N)} \right\vert^2\right]$ dB	$k = 0 \ldots N/2$
where $s(l)$ is the input signal and $j = \sqrt{-1}$	

which requires calculation of $LT_g(i)$, the global masking threshold at the ith frequency sample.

$$LT_g(i) = 10\log_{10}\left[10^{LT_q(i)/10} + \sum_{j=1}^{m} 10^{LT_{tm}[z(i),z(j)]/10} + \sum_{j=1}^{n} 10^{LT_{nm}[z(i),z(j)]/10}\right] \tag{9.8}$$

Thus, $LT_g(i)$ requires the calculation of $LT_q(i)$, the Absolute Threshold, $LT_{tm}[z(j), z(i)]$, the individual Masking Threshold for tonal maskers, and $LT_{nm}[z(j), z(i)]$, the Individual Masking Threshold for nontonal maskers.

Determining the Absolute Threshold

When a tone of a certain amplitude and frequency is present, other tones or noise near this frequency cannot be heard by the human ear. The maximum level of the lower amplitude signal that is imperceptible is called the *Masking Threshold.* There is a threshold, which is a function of frequency, below which a tone or noise is inaudible regardless of whether there is a masker or not. This threshold is called the *Absolute Threshold* and is shown if Figure 9.3.

The *Absolute Threshold* is labeled $LT_q(k)$ and is also called threshold in quiet. In the ISO Standard, Table D.1f gives a set of index numbers, Frequencies for Critical Bands, Critical Band Rates (CBRs), and the Absolute Threshold. The

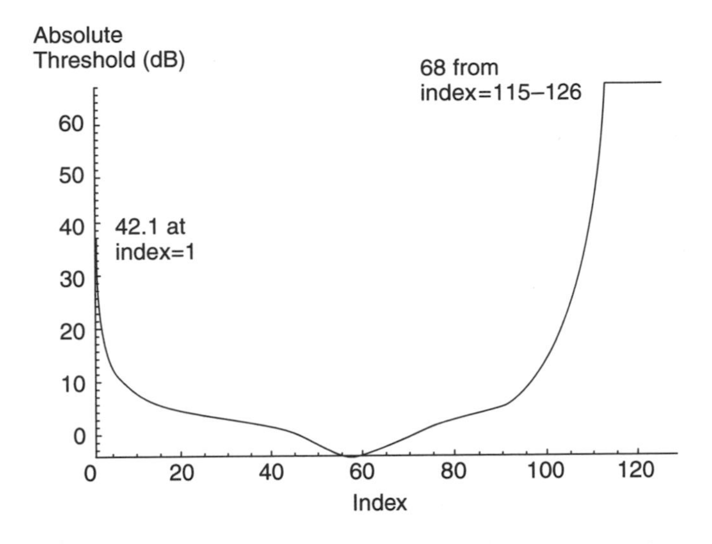

Figure 9.3 Absolute Threshold

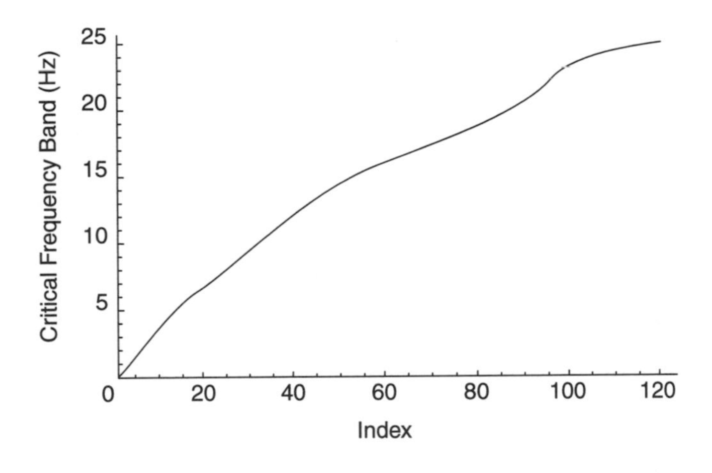

Figure 9.4 Critical Frequency Band

index runs from 1 to 126. For most work, the values of these parameters can be determined from the curves in Figures 9.3, 9.4, and 9.5.

Individual Masking Threshold Determination

Tonal maskers are close to sinusoids and have spectra that are near impulses. On the other hand, nontonal maskers exhibit noiselike characteristics. The

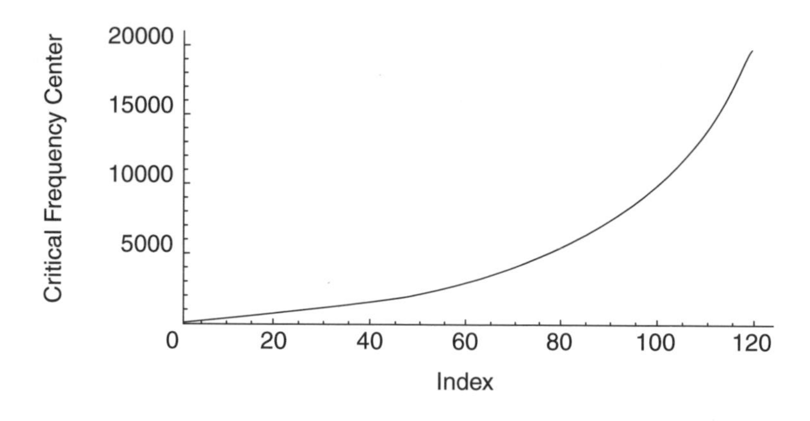

Figure 9.5 Critical Frequency

individual masking thresholds of both tonal and nontonal components are given by the following expressions:

$$LT_{tm}[z(j), z(i)] = X_{tm}[z(j)] + av_{tm}[z(j)] + vf[z(j), z(i)] \text{ dB} \tag{9.9}$$

$$LT_{nm}[z(j), z(i)] = X_{nm}[z(j)] + av_{nm}[z(j)] + vf[z(j), z(i)] \text{ dB} \tag{9.10}$$

In Equations (9.9) and (9.10), LT_{tm} and LT_{nm} are the individual Masking Thresholds at CBR z in Bark of the masking component at the Critical Band Rate of the masker z_m in Bark. The values in dB can be either positive or negative. Equations (9.9) and (9.10) require the calculation of five new parameters: $X_{nm}[z(j)]$, $X_{tm}[z(j)]$, $av_{nm}[z(j)]$, $av_{tm}[z(j)]$, and $vf[z(j), z(i)]$.

Finding of $X_{nm}[z(j)]$ and $X_{tm}[z(j)]$

It is necessary to discriminate between them when calculating the Global Masking Threshold from the FFT spectrum. This step starts with the determination of local maxima, then extracts tonal components and calculates the intensity of the nontonal components within a bandwidth of a critical band. The following describes how boundaries of the critical bands can be determined.

The ISO Standard, in Table D.2f, gives these values, which are just a subset of 27 of the 126 values from Table D.1f. The index runs from 0 to 26, representing the indices 1, 2, 3, 5, 7, 9, 12, 14, 17, 20, 24, 27, 32, 37, 42, 49, 53, 59, 65, 73, 77, 82, 89, 97, 103, 113, and 126. The values for the frequency and band can be determined from Figures 9.4 and 9.5.

The bandwidth of the critical bands varies with the center frequency, with a bandwidth of about only 0.1 kHz at low frequencies and with a bandwidth of about 4 kHz at high frequencies. It is known from psychoacoustic experiments that the ear has a better frequency resolution in the lower than in the higher frequency region. To determine if a local maximum may be a tonal component, a frequency range, df, around the local maximum is examined. For frequencies up to 3 kHz, df = 93.75 Hz. For frequencies from 3 kHz to 6 kHz, df = 140.63 Hz. For frequencies from 6 kHz to 12 kHz, df = 281.25 Hz, and for frequencies between 12 kHz and 24 kHz, df = 562.5 Hz.

To make lists of the spectral lines $X(k)$ that are tonal or nontonal, the following five operations are performed:

1. *Selection of local maxima candidates*—A spectral line $X(k)$ is labeled as a local maximum if

$$X(k) > X(k-1) \text{ and } X(k) >= X(k+1) \tag{9.11}$$

2. *Selection of tonal components*—A local maximum is put in the list of tonal components if

$$X(k) - X(k+j) >= 7 \text{ dB} \tag{9.12}$$

where j is chosen according to the subband index range. For the index between 2 and 63, $j = \pm 2$; for the index greater than or equal to 63 and less than 127, $j = \pm 2, \pm 3$; for index greater than or equal to 127 and less than 255, $j = \pm 2, \pm 3, \pm 4, \pm 5$, and ± 6. This step checks how wide the spectrum of the local maximum candidates is.

3. *Sound pressure level*—The SPL is calculated next. For each k, where $X(k)$ has been found to be a tonal component,

$$X_{tm}(k) = 10 * \log_{10}\left[10^{\frac{X(k-1)}{10}} + 10^{\frac{X(k)}{10}} + 10^{\frac{X(k+1)}{10}}\right], \text{ in dB} \tag{9.13}$$

Next, all spectral lines within the examined frequency range are set to $-\infty$ dB.

4. *Nontonal components*—The nontonal (noise) components are calculated from the remaining spectral lines. To calculate the nontonal components from spectral lines $X(k)$, the critical bands $z(k)$ are determined using Figure 9.5. In Layer II, 26 critical bands are used for the 48 kHz sampling rate.

5. *Calculation of the power*—Within each critical band, the power of the spectral lines (remaining after the tonal components have been zeroed) are summed to form the SPL of the new nontonal component, $X_{nm}(k)$, corresponding to that critical band.

Decimation of Tonal and Nontonal Masking Components

Decimation is a procedure that is used to reduce the number of maskers that are considered for the calculation of the Global Masking Threshold. In the discussion here, a Bark is the critical band unit.

> $X_{tm}(k)$ or $X_{nm}(k)$ are considered for the calculation of the Masking Threshold only if they are greater than $LT_q(k)$ for each index k.
>
> Decimation of two or more tonal components within a distance of less than 0.5 Bark: Keep the $X_{tm}[z(j)]$ with the highest power, and remove the smaller component(s) from the list of tonal components. For this operation, a sliding window in the critical band domain is used with a width of 0.5 Bark.

The term av is called the masking index and is different for tonal and nontonal maskers (av_{tm} and av_{nm}). The masking indices av_{tm} and av_{nm} are empirically determined. For tonal maskers it is given by

$$av_{tm} = -1.525 - 0.275 * z(j) - 4.5 \text{ dB} \quad (9.14)$$

and for nontonal maskers by

$$av_{tm} = -1.525 - 0.175 * z(j) - 0.5 \text{ dB} \quad (9.15)$$

The term vf is called the masking function of the masking component $X_{tm}[z(j)]$ and is characterized by different lower and upper slopes, which depend on the distance in Bark, $dz = z(i) - z(j)$, to the masker. In this expression, i is the index of the spectral line at which the masking function is calculated and j is that of the masker. CBRs $z(j)$ and $z(i)$ can be found in Figure 9.4.

The masking function is defined in terms of dz and $X[z(j)]$. It is a piecewise linear function of dz, nonzero for $-3 \le dz < 8$ Bark. The parameter vf depends only weakly on $X[z(j)]$. This can be seen in Figure 9.6. LT_{tm} and LT_{nm} are set to $-\infty$ dB if $dz < -3$ Bark, or $dz >= 8$ Bark.

9.3.3 Bit Allocation

The process of allocating bits to the coded parameters must attempt to simultaneously meet both the bit rate requirements and the masking requirements. If insufficient bits are available, it must use the bits at its disposal in order to be as psychoacoustically inoffensive as possible. In Layer II, this method is a bit allocation process (i.e., a number of bits are assigned to each sample, or group of samples, in each subband).

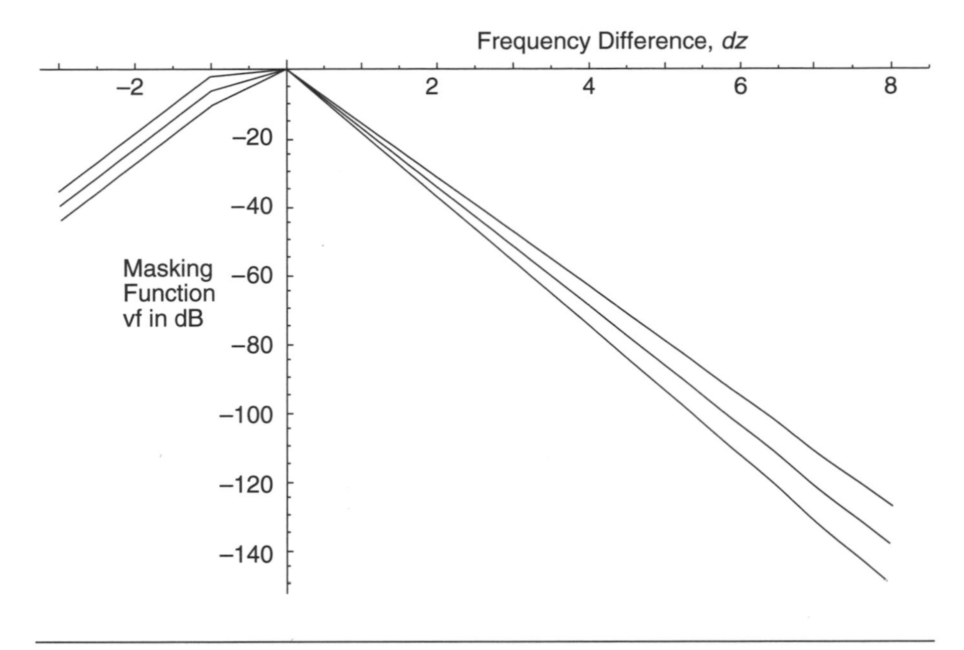

Figure 9.6 Masking Function for $X(z(j)) = -10, 0, 10$

Scalefactor Calculation

The scalefactors can be stored in a table (see Table B.1 in the ISO Standard) or calculated. Using the number 1.2599210498949, the scalefactor value for each index n is

$$\frac{2.00000000000000}{1.2599210498949^n} \tag{9.16}$$

where n ranges from 0 to 62. For each subband, a scalefactor is calculated for every 12 subband samples. The maximum of the absolute value of these 12 samples is determined. The lowest value calculated from Equation (9.16) that is larger than this maximum is used as the scalefactor.

Coding of Scalefactors

Call the index determined from the algorithm of the Possible Quantization per Subband section scf. Because there are three of these per subband, they are labeled scf_1, scf_2, and scf_3. First, the two differences, $dscf_1$ and $dscf_2$, of the successive scalefactor indices—scf_1, scf_2, and scf_3—are calculated:

$$dscf_1 = scf_1 - scf_2 \tag{9.17}$$

$$dscf_2 = scf_1 - scf_3 \tag{9.18}$$

The class of each of the differences is determined as follows:

- If dscf is less than or equal to –3, the class is 1.
- If dscf is between –3 and 0, the class is 2.
- If dscf equals 0, the class is 3.
- If dscf is greater than 0 but less than 3, the class is 4.
- If dscf is greater than 3, the class is 5.

Because there are two dscf, there are two sets of classes.

The pair of classes of differences indicates the entry point in the following algorithm to get the calculations in ISO Table C.4, "Layer II Scalefactor Transmission Patterns." Because there are two sets of five classes, there are 25 possibilities:

- For the class pairs (1,1), (1,5), (4,5), (5,1), and (5,5),[1] all three scalefactors are used, the transmission pattern is 123, and the Selection Information is '00'.
- For the class pairs (1,2), (1,3), (5,2), and (5,3), the scalefactors used are 122, the transmission pattern is 12, and the Selection Information is '11'.
- For the class pairs (1,4) and (5,4), the scalefactors are 133, the transmission pattern is 13, and the Selection Information is '11'.
- For the class pairs (2,1), (2,5), and (3,5), the scalefactors used are 113, the transmission pattern is 13, and the Selection Information is '01'.
- For classes (2,2), (2,3), (3,1), (3,2), and (3,3), the scalefactors are 111, the transmission pattern is 1, and the Selection Information is '01'.
- For class pairs (3,4) and (4,4), the scalefactors are 333, the transmission pattern is 3, and the Selection Information is '01'.
- If the class pairs are (4,1), (4,2), and (4,3), the scalefactors are 222, and the Selection Information is '01'.
- Finally, class pair (2,4) uses the scalefactors 444, the transmission pattern 4, and Selection Information '01'.

---

[1] "1," "2," and "3" mean the first, second, and third scalefactor within a frame, and "4" means the maximum of the three scalefactors. The information describing the number and the position of the scalefactors in each subband is called "Scalefactor Selection Information."

Scalefactor Selection Information Coding

For the subbands that will get a nonzero bit allocation, the Scalefactor Selection Information (scfsi) is coded by a 2-bit word. This was noted previously for Layer II Scalefactor Transmission Patterns.

Bit Allocation

Before starting the bit allocation process, the number of bits, "adb", that are available for coding the payload must be determined. This number can be obtained by subtracting from the total number of available bits, "cb"; the number of bits needed for the header, "bhdr" (32 bits); the CRC checkword, "bcrc" (if used) (16 bits); the bit allocation, "bbal"; and the number of bits, "banc," required for ancillary data:

$$\text{adb} = \text{cb} - (\text{bhdr} + \text{bcrc} + \text{bbal} + \text{banc}) \tag{9.19}$$

Thus, adb bits are available to code the payload subband samples and scalefactors. The allocation approach is to minimize the total noise-to-mask ratio over the frame.

Possible Quantization per Subband: This data can be presented as a table (B.2a in the ISO Standard) or computed from the following:

1) The first 3 subbands (0, 1, 2) have 15 possible values with indices from 1 to 15. The values can be calculated from $2^{i+1} - 1$, where i is the index.
2) The next 8 subbands (3–10) also have 15 possible values. For index 1 to 4, the values are 3, 5, 7, and 9. For index 5–14, the entry is 2^{i-1}. For index = 15, the value is 65,535.
3) Subbands 11–22 have only 7 indices. For indices 1–6, the values are 3, 5, 7, 9, 15, and 31. The value for index 7 is 65,535.
4) Subbands 23–26 have only 3 indices. The values are 3, 5, and 65,535.
5) Subbands 27–31 have no quantization value.

The parameter nbal indicates the number of bits required to determine the number of populated indices. Thus, nbal = 4 for the first 10 subbands, 3 for subbands 11–22, and 2 for subbands 22–26. The number of bits required to represent these quantized samples can be derived from the following algorithm for "Layer II Classes of Quantization." The allocation is an iterative procedure. The rule is to determine the subband that will benefit the most and increase the bits allocated to that subband.

For each subband, the mask-to-noise ratio (MNR) is calculated by subtracting from the signal-to-noise ratio (SNR) the signal-to-mask ratio (SMR). The minimum MNR is computed for all subbands.

$$\mathrm{MNR} = \mathrm{SNR} - \mathrm{SMR} \tag{9.20}$$

Because these entities are in dB, this is equivalent to dividing SNR by SMR; hence the name MNR, since

$$\frac{\frac{S}{N}}{\frac{S}{M}} = \frac{M}{N} = \mathrm{MNR}$$

The SNR can be found from Figures 9.7 and 9.8. The SMR is the output of the psychoacoustic model. Because step size is growing exponentially, Figure 9.7 has a logarithmic shape that doesn't show the low values very well. Figure 9.8 shows just the low values of step size.

The number of bits for (1) the samples "bspl", (2) scalefactors "bscf", and (3) Scalefactor Selection Information "bsel" are set to zero. The minimum MNR is increased by using the next higher index in the previous Possible Quantization per Subband section and a new MNR of this subband is calculated.

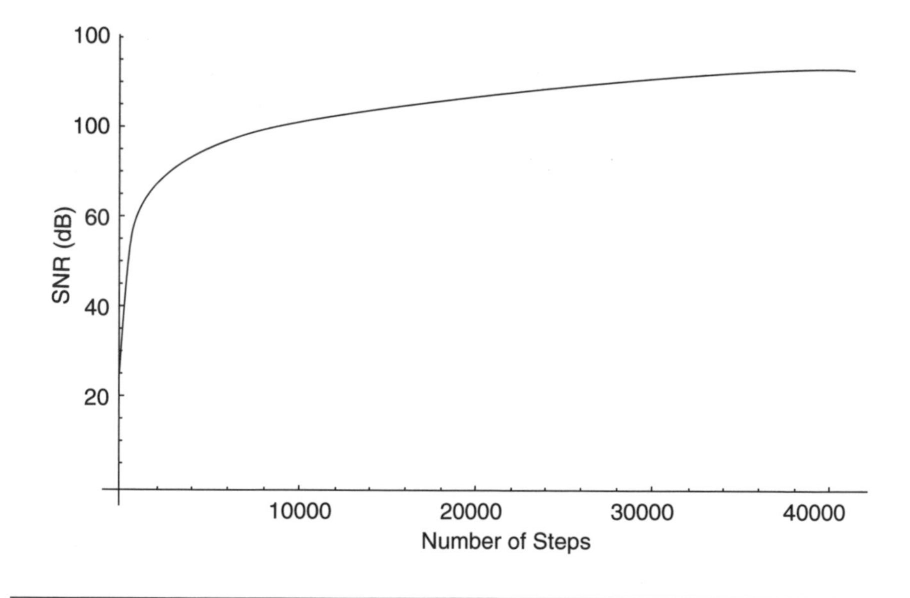

Figure 9.7 Signal-to-Noise versus Step Size

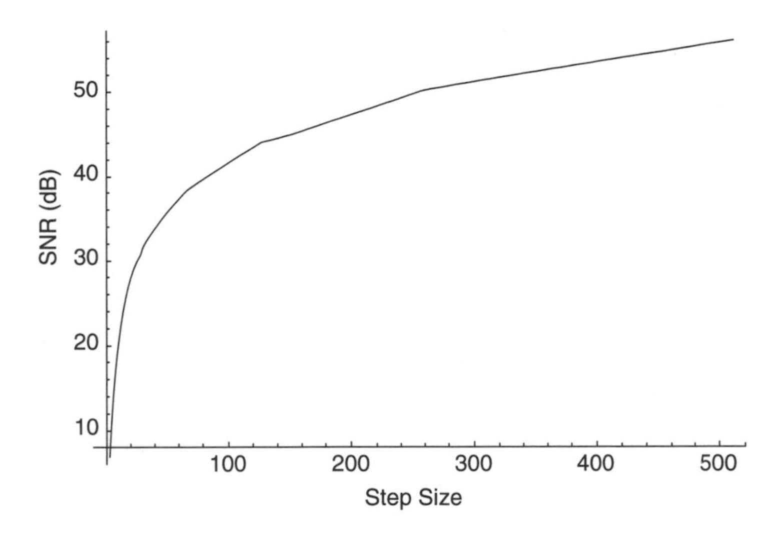

Figure 9.8 Signal-to-Noise versus Step Size (for Small Step Sizes)

The number of bits for samples, bspl, is updated according to the additional number of bits required. If a nonzero number of bits is assigned to a subband for the first time, bsel and bscf have to be updated. Then adb is calculated again using the formula:

$$adb_{next} = cb - (bhdr + bcrc + bbal + bsel + bscf + bspl + banc) \qquad (9.21)$$

The iterative procedure is repeated as long as adb_{next} is not less than any possible increase of bspl, bsel, and bscf within one loop.

Quantization and Encoding of Subband Samples

The quantized sample $Q(i)$ is calculated by

$$Q(i) = A * X(i) + B \qquad (9.22)$$

where the $X(i)$ are the subband samples. A and B are determined with the following algorithm:

1. There are 17 indices. The first 14 are the same as noted previously under the Possible Quantization section. For indices 15–17, the value is $2^{i-1} - 1$.
2. The A values for an index are the index value divided by the minimum power of 2 that is larger than the index value. For example, take index

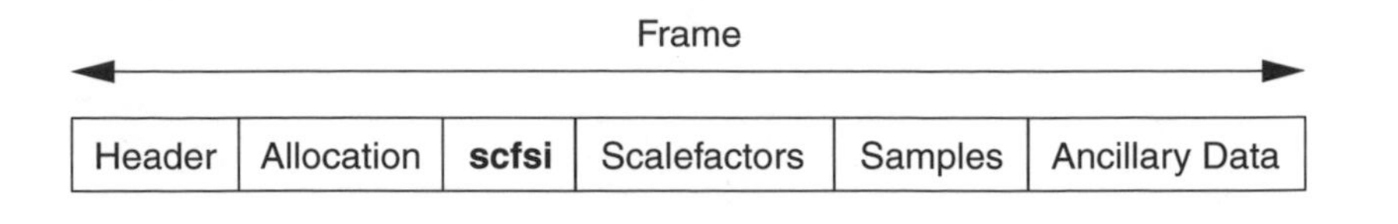

Figure 9.9 Layer II Packet Format Adapted from ISO Standard 11172-3, Figure C.3.

value 9. The next higher power of 2 is 16. The *A* value is then 9/16 = .5625.

3. The *B* values are all negative. The values can be calculated as follows: The denominator is $2^{\text{number of bits to represent the index}}$. The numerator is the denominator minus the index. For example, consider index 9. The bits required to represent 9 are 4 and $2^4 = 16$. Thus, the denominator is 16. The numerator is then 16 − 9 = 7, and the value is −7/16 = −.4375.

N represents the necessary number of bits to encode the number of steps. The inversion of the MSB is done in order to avoid the all '1' code that is used for the synchronization word.

The algorithm of the Possible Quantization section on page 187 shows whether grouping is used. The groups of three samples are coded with individual codewords if grouping is not required.

The three consecutive samples are coded as one codeword if grouping is required. Only one value v_m, MSB first, is transmitted for this triplet. The empirical rule for determining v_m (m = 3, 5, 9) from the three consecutive subband samples x, y, z is:

$$v_3 = 9z + 3y + x \quad (v_3 \text{ in } 0 \ldots 26) \tag{9.23}$$

$$v_5 = 25z + 5y + x \quad (v_5 \text{ in } 0 \ldots 124) \tag{9.24}$$

$$v_9 = 81_z + 9y + x \quad (v_9 \text{ in } 0 \ldots 728) \tag{9.25}$$

9.3.4 Formatting

The formatting for Layer II packets is given in Figure 9.9.

9.4 The Audio Decoding Process

To start the decoding process, the decoder must synchronize itself to the incoming bitstream. Just after startup this may be done by searching in the bitstream for the 12-bit syncword.

9.4.1 General

In DBS a number of the bits of the header are already known to the decoder, and, thus, can be regarded as an extension of the syncword, thereby allowing a more reliable synchronization. For Layer II, the number of bytes in a packet is

$$\mathrm{N} = 144 * \frac{\text{bit rate}}{\textbf{sampling\_frequency}} \text{ bytes} \tag{9.26}$$

For DBS,

$$N = 144 * \frac{192,000}{48,000} = 576 \text{ bytes} = 4,608 \text{ bits} \tag{9.27}$$

For DBS, the mode bits are '01' for the **intensity_stereo** and the **mode_extension** bits apply. The **mode_extension** bits set the **bound,** which indicates which subbands are coded in **joint_stereo** mode.

A CRC checkword has been inserted in the bitstream if the **protection_bit** in the header equals '0'.

9.4.2 Layer II Decoding

The Audio decoder first unpacks the bitstream and sends any ancillary data as one output. The remainder of the output goes to a reconstruction function and then to an inverse mapping that creates the output left and right audio samples.

Audio Packet Decoding

The parsing of the Audio packet is done by using the following three-step approach.

1. The first step consists of reading "nbal" (2, 3, or 4) bits of information for one subband from the bitstream. These bits are interpreted as an unsigned integer number.
2. The parameter nbal and the number of the subband are used as indices to compute nlevels that were used to quantize the samples in the subband.
3. The number of bits used to code the quantized samples, the requantization coefficients, and whether the codes for three consecutive subband samples have been grouped into one code can be determined.

The identifier **sblimit** indicates the number of the lowest subband that will not have bits allocated to it.

Scalefactor Selection Information Decoding

There are three equal parts (0, 1, 2) of 12 subband samples each. Each part can have its own scalefactor. The number of scalefactors that has to be read from the bitstream depends on **scfsi[sb].** The Scalefactor Selection Information **scfsi[sb]** decodes the encoding shown in the Coding of Scalefactors section earlier in this chapter.

Scalefactor Decoding

The parameter **scfsi[sb]** determines the number of coded scalefactors and the part of the subband samples they refer to. The 6 bits of a coded scalefactor should be interpreted as an unsigned integer index. The Possible Quantization of Subbands section on page 187 gives the scalefactor by which the relevant subband samples should be multiplied after requantization.

Requantization of Subband Samples

The coded samples appear as triplets. From the Possible Quantization section, it is known how many bits have to be read for each triplet for each subband. As determined there, it is known whether the code consists of three consecutive separable codes for each sample or of one combined code for the three samples (grouping). If the encoding has employed grouping, degrouping must be performed. If combined, consider the word read from the bitstream as an unsigned integer, called c. The three separate codes—$s[0]$, $s[1]$, $s[2]$—are calculated as follows:

$$
\begin{aligned}
s[0] &= c \text{ modulo[nlevels]}^{2} \\
c[1] &= c \text{ DIV nlevels} \\
s[1] &= c[1] \text{ modulo[nlevels]} \\
c[2] &= c[1] \text{ DIV nlevels} \\
s[2] &= c[2] \text{ modulo[nlevels]}
\end{aligned}
\tag{9.28}
$$

To reverse the inverting of the MSB done in the encoding process, the first bit of each of the three codes has to be inverted. The requantized values are calculated by

$$s'' = C * (s''' + D) \tag{9.29}$$

---

[2] *Note:* (1) DIV is defined as integer division with truncation towards; and (2) nlevels is the number of steps, as shown in the Possible Quantization per Subband section.

where

$$s''' = \text{the fractional number}$$

$$s'' = \text{the requantized value}$$

To calculate the coefficient C (not the C_i used in the Analysis Filterbank), let n equal the number of steps. Then,

$$C = \frac{2^j}{n}$$

where j is the smallest number of bits that can represent n. For example, if $n = 9$, then $j = 4$, since 4 bits are required to represent 9. Then,

$$C = \frac{2^4}{9} = \frac{16}{9} = 1.77777$$

The parameter D (not the D_i used in the Synthesis Filterbank) can be calculated by

$$D = \frac{1}{2^{j-1}}$$

where j is the number of bits required to represent n. For example, if $n = 31$, $j = 5$, and

$$D = \frac{1}{2^4} = .0625$$

If the number of steps is 5 or 9, $D = .5$. If the number of steps is 3, 5, or 9, there is a grouping and the samples per codeword is 3; otherwise, there is no grouping and the samples per codeword is 1. The bits per codeword equals j, except where the number of steps is 3, 5, or 9, where the bits per codeword are 5, 7, and 10, respectively.

The parameters C and D are used in Equation (9.28). The requantized values have to be rescaled. The multiplication factors can be found in the calculation of scalefactors as described previously. The rescaled value s' is calculated as:

$$s' = \text{factor} * s'' \tag{9.30}$$

Synthesis Subband Filter

If a subband has no bits allocated to it, the samples in that subband are set to zero. Every time the subband samples for all 32 subbands of one channel have been calculated they can be applied to the synthesis subband filter and 32 consecutive audio samples can be calculated. For that purpose, the actions shown in the following algorithm detail the reconstruction operation.

The coefficients N_{ik} for the matrixing operation are given by

$$N_{ik} = \cos\left[(16+i)*(2k+1)*\frac{\pi}{64}\right], \quad 0 \le i \le 63, \; 0 \le k \le 31 \qquad (9.31)$$

The coefficients D_i for the windowing operation can be found from Figure 9.2. One frame contains 36 * 32 = 1,152 subband samples, which create (after filtering) 1,152 audio samples.

Synthesis Filterbank Algorithm

1. Input 32 new subband samples
2. Using these samples, perform the matrix multiply

$$\begin{bmatrix} V_0 \\ V_1 \\ \cdot \\ \cdot \\ \cdot \\ V_{63} \end{bmatrix} = \begin{bmatrix} N_{00} & N_{01} & \cdot & \cdot & \cdot & N_{0,31} \\ N_{10} & \cdot & \cdot & \cdot & \cdot & N_{1,31} \\ \cdot & \cdot & \cdot & \cdot & \cdot & \cdot \\ \cdot & \cdot & \cdot & \cdot & \cdot & \cdot \\ \cdot & \cdot & \cdot & \cdot & \cdot & \cdot \\ N_{63,0} & \cdot & \cdot & \cdot & \cdot & N_{63,31} \end{bmatrix} * \begin{bmatrix} F_0 \\ F_1 \\ \cdot \\ \cdot \\ \cdot \\ F_{31} \end{bmatrix}$$

Sixteen of these V vectors are shifted into the FIFO shown here. Perform V into U mapping.

0	64	128	192	256	320	384	448	512	578	640	704	768	832	896	960
X		**X**		**X**		**X**		**X**		**X**		**X**		**X**	
31	95	159	223	287	351	415	479	543	607	671	735	799	863	927	991
32	96	160	224	288	352	416	480	544	608	672	736	800	864	928	992
	X		**X**		**X**		**X**		**X**		**X**		**X**		**X**
63	127	191	255	319	383	447	511	575	639	703	767	831	895	959	1023

X indicates where the V FIFO maps into U

4. Window the U vector into a W vector by

$$W_i = D_i * U_i \quad i = 0, 1, \ldots, 511 \tag{9.32}$$

5. Calculate the Output Audio samples

$$S_j = \sum_{l=0}^{15} W_{j+32*i} \quad j = 0 \text{ to } 31 \tag{9.33}$$

PART FOUR

Ground Subsystems

CHAPTER TEN

DBS Uplink Facilities

The DBS uplink was introduced in Chapter 3. In this chapter, the components of the uplink facility are discussed in more detail.

The DBS uplink facilities are very complicated electronic systems. The broadcast facilities for a transponder are roughly equivalent to those of a television station. Since the typical uplink facility provides for 20 to 30 transponders, it has the complexity of more than 20 television stations. What's more, the uplink must continuously provide for 200 television channels. Each of these must be compressed and time-division multiplexed (TDM) into bitstreams for transmission to the satellite transponders.

Another way to view the complexity of a DBS uplink facility is to examine the bit rates involved. When a DBS orbital slot is fully populated, 32 transponders will be operating at 30.3 Mbps (information bit rate) for a total of 970 Mbps. Since the compression ratios are 40–60:1, 40 to 60 gigabits per second (or the equivalent) must be fed into the system every second of every day.

10.1 Existing Uplink Facilities

As of December 1996, there were four operating uplink facilities, as shown in Table 10.1. Figure 10.1 is a photograph of the DIRECTV uplink facility at Castle Rock, Colorado.

Table 10.1 Operating Uplink Facilities

Company	Location
DIRECTV	Castle Rock, Colorado
U.S. Satellite Broadcasting	Oakdale, Minnesota
EchoStar	Cheyenne, Wyoming
AlphaStar	Oxford, Connecticut

Figure 10.1 Castle Rock, Colorado, Uplink Facility. From Jim C. Williams, "Airborne Satellite Television," ***Avion***, Fourth Quarter:44, 1994. Used with permission.

Organization of uplink facilities by transponder (or group of transponders) would solve the TDM versus FDM issue; that is, the system would be FDM by transponder and TDM within a transponder. However, the investment in common features, such as broadcast automation, plus the issues of synchronizing a common program guide and so on, have led all the companies listed in Table 10.1 to have a single uplink facility for all their transponders.

10.2 Constituents of an Uplink Facility

The following information is excerpted from a paper presented at the 30th SMPTE Conference in February 1996 [Echeita96]. The equipment in an uplink facility includes:

- A fully automated Sony broadcast system for 200 video and audio channels
- A 512-by-512 output Sony digital routing system that carries four audio signals for each video signal
- Fifty-six Sony Flexicart robotic videotape playback systems
- Two 1,000-cassette Sony Library management systems
- More than 300 Sony Digital Betacam videotape machines
- State-of-the-art scheduling hardware and software
- Fifty-six Multichannel Compression Systems (see next section)
- Four 13-meter transmitting antennas plus multiple receive antennas

The uplink facility includes, among other things, the following:

- Three 1.5-megawatt generators
- An Uninterruptible Power System (UPS)
- A 100,000-gallon fire-suppression system
- Computerized environmental monitoring and control

10.3 Key Uplink Subsystems

Much has been written about uplink facilities and equipment. The two subsystems of particular interest for digital communications are the compressor/multiplexer and the QPSK modulator.

10.3.1 Compressor/Multiplexer

Several companies develop and manufacture uplink equipment, including Compression Laboratories Inc. (CLI), a division of General Instrument (GI), and DiviCom. The following discussion emphasizes the CLI/GI system called the Magnitude™ [CLI96] solely because the author is more familiar with it.

Figure 10.2 is a block diagram of a compression/multiplexer system. The multiplexer block contains the common elements, including the mounting rack, power supplies, and so forth. The blocks marked video compressors

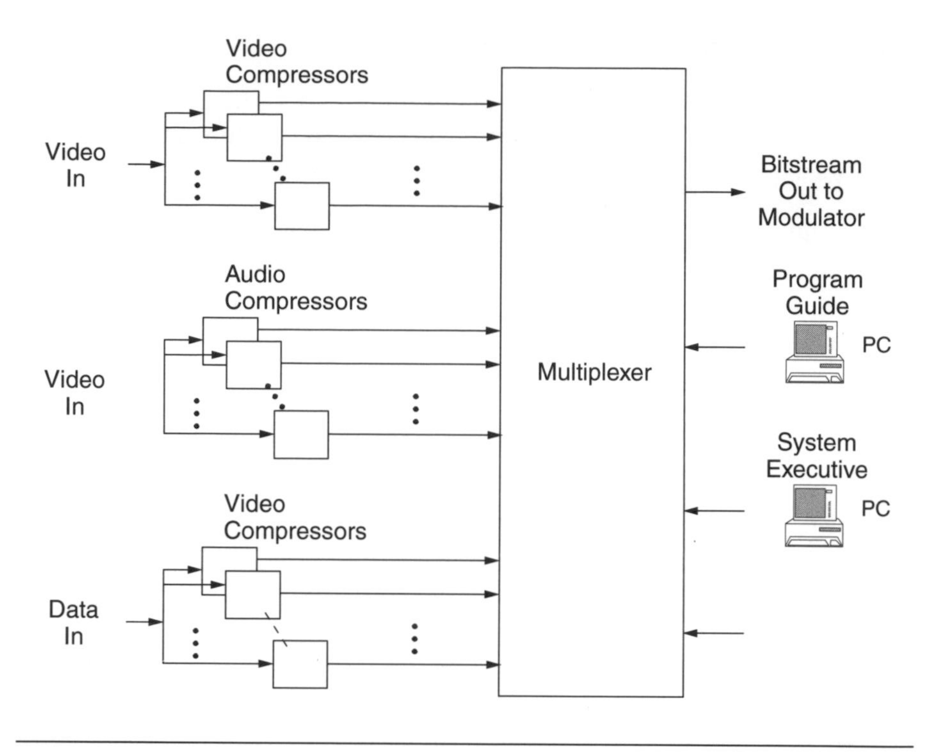

Figure 10.2 Compressor-Multiplexer Block Diagram

accept the incoming video formats (S-Video or D-1 are preferred) and compress the service according to MPEG 2 MP@ML as described in Chapter 8. There is no theoretical limit to the number of such video compressors in a single multiplex, although 6 to 8 is a practical range.

The audio encoders accept digital audio in the form of 48 Ksamples per second, 16 bits per sample, and compress it according to MPEG 1 (ISO 11172-3) Layer II at 192 Kbps (Chapter 9). The encoders offer a range of bit rates but 192 Kbps is used by all known DBS systems.

User data also can be an input to the multiplexer in compliance with MPEG 2 Systems (ISO 13818-1, 1994), as discussed in Chapter 7. Data may come in many forms. In Chapter 6, it was shown that the delivery of ECMs and EMMs, along with other authentication and signature data, must be transmitted as User Data. Similarly, Electronic Program Guide data must be sent as other data.

One interesting use of User Data involves audio services. When MPEG 2 Audio (ISO 13818-3) established its six-channel audio, it did so in a backward-compatible way. This means that an MPEG 1-compliant decoder can derive a stereo audio from an MPEG 2 audio bitstream. This is accomplished by coding the left front and right front audio inputs as the left and right inputs from MPEG 1. The center, right surround, left surround, and subwoofer channels are compressed and sent as User Data.

An MPEG 1 audio decoder decodes the left and right signals to create normal stereo. A more sophisticated receiver can use the User Data plus the audio data to create a six-channel sound for a home theater. Note that such an implementation can be done without making an installed base obsolete. It is only an encoder change.

The Conditional Access input provides encryption keys for the Multiplexer, which applies them to the service streams as well as to other data that will be used by IRDs for accessing services.

The Electronic Program Guide is provided by another PC that provides its information as User Data with a specific PID (see Chapter 7). The current time resolution of Electronic Program Guides is 30 minutes, although this probably will decrease in the future.

The System Executive PC performs a number of functions, including overall supervision of the system. Through it, various services are assigned PIDs. It also manages the statistical multiplexing and system clock distribution.

10.3.2 The Modulator

To call this unit a modulator is a serious understatement. Both of the systems with which I am familiar use Comstream modulators: CM2000 for DSS waveform systems and CM3000 for DVB systems. The CM2000 is discussed in this section.

The functions of the Reed-Solomon Coder/Interleaver/Convolutional Coder were covered in Chapter 5. Energy dispersal, raised cosine filter, and QPSK modulator were discussed in Chapter 3. Figure 10.3 is a simplified diagram of the modulator that shows how all these functions come together. They will not be dealt with further here. Instead, the role of the frequency synthesizer and its interaction with the multiplexer will be explained.

For DIRECTV, the frequency synthesizer puts out a frequency 2/3 (40) = 26.67 Mbps or 6/7 (40) = 34.29 Mbps, depending on which code rate is being employed. DVB outputs comparable bit rates for its code rates. This clock is used by the multiplexer to clock out the serial bitstream. If an external 10-MHz clock is present, the frequency synthesizer will lock to this reference (to the nearest integer symbol).

For DIRECTV, the compressor/multiplexer organizes the data into packets of 1,176 bits each (147 * 8). These consist of 1,040 bits of information, 128 bits of stuffed zeros where the Reed-Solomon parity bits will be added,

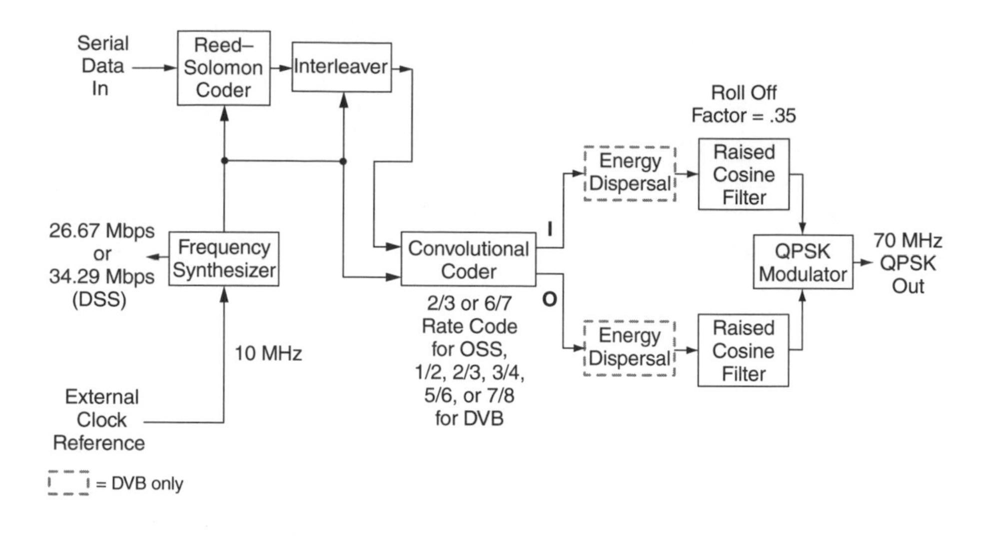

Figure 10.3 Modulator Block Diagram

and 8 bits of stuffed zeros where a sync word will be added. The sequence of 1,176 bits repeats every 44.1 seconds for the low-rate code and every 34.3 seconds for the high-rate code.

The modulator receives the 1,176 bits, inserts 128 Reed-Solomon parity bits in place of the stuffed zeros, and adds an 8-bit sync word. It then adds the convolutional code bits (at a code rate of 2/3 or 6/7) and outputs the constant 40-Mbps serial bitstream to the QPSK modulator. Note that the 40-Mbps rate is constant regardless of code rate. Because the QPSK modulator codes two bits per symbol, the symbol rate is 20 mega symbols per second.

10.4 Statistical Multiplexing

All known DBS systems use time-division multiplexing within a transponder. Further, natural video is very bursty in nature. In compressing CCIR 601-1 video, MPEG 2 can require as low as 1 Mbps and as high as 10 Mbps to create good-quality encoded video.

Thus, a dilemma exists. If the bit rate for a service is set at the minimum required to give good quality, say 10 Mbps, this will be very wasteful because this rate will be required only a small fraction of the time. On the other hand, if a low rate is set, there will be times when the quality is too low.

The solution to this dilemma is called statistical multiplexing. Recall from Chapter 8 that MPEG 2 video compressors have a rate-control mechanism.

Essentially, this consists of controlling the quantization levels within certain limits so that the buffers stay within certain limits.

If this control (for all services within a multiplexer) is given to the multiplexer, it can allocate bits to each service according to its requirements. Thus, the individual compressors become variable bit rate, constant quality.

What happens if all (or a large number) of the services within a multiplex all burst up simultaneously? This will happen, although infrequently. In this case, all the services will have to suffer some temporary quality degradation. This usually will be of such short duration that the average viewer will not even notice.

Statistical multiplexing is an active area of research, and advancements are still being made.

References

[Bursky96] Bursky, Dave, Full Feature MPEG 2 Encoder Needs Just Five Logic ICs, *Electronic Design*, April 1996.

[CITEL94] CITEL Inter-American Telecommunication Commission, *Final Report*, First Meeting of Permanent Consultative Committee II: Radiobroadcasting (PCC.II), August 1994.

[CLI96] Magnitude Technical White Paper, Compression Laboratories Inc., Internet Document, 1996.

[Echeita96] Echeita, Richard J., Challenges in a Digital, Server-Based Broadcasting Environment. In *Proceedings of the 30th SMPTE Advanced Motion Imaging Conference* (Session 5), February 1996.

[Williams94] Williams, Jim C., Technologies for Direct Broadcast to Airliners, *AVION* (Fourth Quarter), 1994.

CHAPTER ELEVEN

Integrated Receiver Decoder

The key to a successful DBS service is an inexpensive integrated receiver decoder (IRD). After all, if consumers cannot afford the IRD, they cannot subscribe to the DBS service! This was considered so important that the first contract for IRDs required that the retail price be $700 or less.

The inexorable march of Moore's law, coupled with the learning curve and production volumes in excess of a million units a year, drove the IRD price down to zero (with one service provider's $200 rebate) as of July 1997.

11.1 IRD Components

The IRD consists of an outdoor unit (ODU) and an indoor unit (IDU). These two units are connected by a coaxial cable. Figure 11.1 is from a photograph of an ODU, which consists of a reflecting antenna plus an LNB (see section 4.4

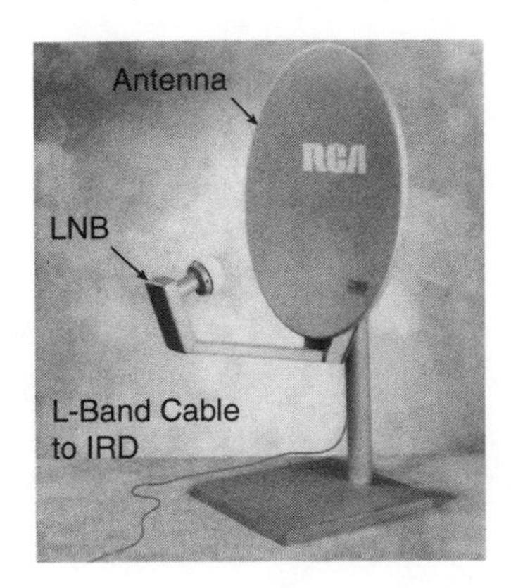

Figure 11.1 IRD Outdoor Unit

Figure 11.2 IRD with Remote Control

in Chapter 4). Note that the LNB is offset from the focus of the parabaloid reflector, thus minimizing the blockage of the incoming signal.

Two types of LNBs are commercially available. One has a single coax connector. The center conductor carries the received signal from the IDU back to the LNB in addition to a DC voltage of either 13 volts or 17 volts. Based on which channel is selected, the IDU determines which of the two voltages to put on the cable, which then selects the proper polarization for the channel.

The second type of LNB has two coax connectors. In the simplest case, the dual LNB can service two IDUs. With some additional components, however, the dual LNB can be used to service a large number of IDUs (see section 11.5).

Figure 11.2 is from a photograph of a typical IDU, with its remote control on top. Figure 11.3 is a photograph of a typical remote control. Both the remote control and the IRD are designed to be especially user-friendly.

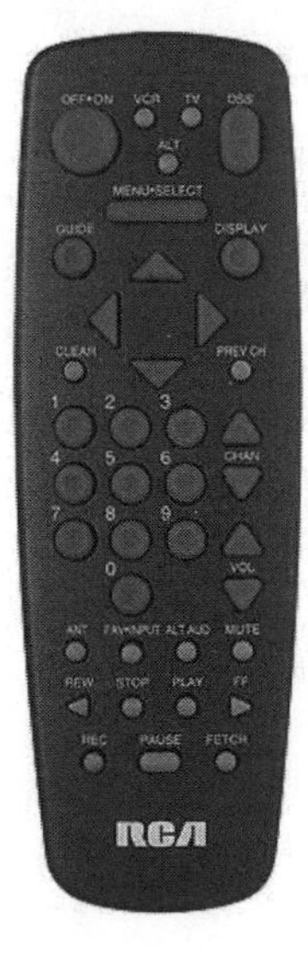

Figure 11.3 Typical Remote Control

11.2 The IRD Architecture

The LNB architecture is shown in section 4.4 , so this section discusses the IDU. Figure 11.4 is a block diagram of the IDU. The L-Band signal from the ODU is the input to the tuner.

The tuner selects one of the 32 transponder signals for further processing. A 5-bit select signal from the microprocessor controller (MC) determines this selection. The tuner is a voltage-controlled oscillator (or a frequency synthesizer) that is called the local oscillator (LO). This is mixed with the incoming L-Band signal to form sum-and-difference frequencies.

The LO frequencies are set so that the difference between the selected transponder center frequency is 12 MHz. Thus, the transponder output occupies the band 0 MHz to 24 MHz and is isolated by a low-pass filter. Another way to view this is that the tuner forms a second heterodyne step: The LNB steps the signal down from K_u-Band to L-Band, and the tuner steps the signal down to low-passed 24 MHz. Of course, the tuner is variable in frequency, so different parts of the L-Band signal can be selected.

The tuner output is input to the demodulator. The QPSK demodulator converts the incoming signal to digital: 2 bits per incoming symbol. The bitstream from the demodulator becomes the input to the decoders of the concatenated code, shown as FEC^{-1} in Figure 11.4.

First, the inverse of the Energy Dispersal is performed. Then, the Viterbi decoder decodes the convolutional code (see section 5.6 in Chapter 5). The Viterbi decoder is followed by the deinterleaver. Finally, the Reed-Solomon decoder creates the original information bitstream.

The first generation of IRDs located the tuner, demodulator, and inverse forward-error-correction circuitry on a daughterboard that is connected to the IRD motherboard. The signal flow was byte parallel from the output of the Reed-Solomon decoder to the motherboard.

The information bytes from the FEC^{-1} function are time-division demultiplexed into individual service byte streams. Note that the selection is one of *n*, where *n* is the number of services that have been time-division multiplexed together on the selected transponder. The output of the demultiplexer is stored in buffer memories until it can be used by the appropriate decryptors and decoders, which provide the output for display.

Control of the IDU is provided by the microprocessor controller (MC). It receives user inputs from the front panel or remote control. Based on these inputs, the MC selects the desired channel for decompression (via the tuner and demultiplexer) and provides the conditional access (CA) controller with this information. The CA controller then provides the correct decryption keys for the selected services.

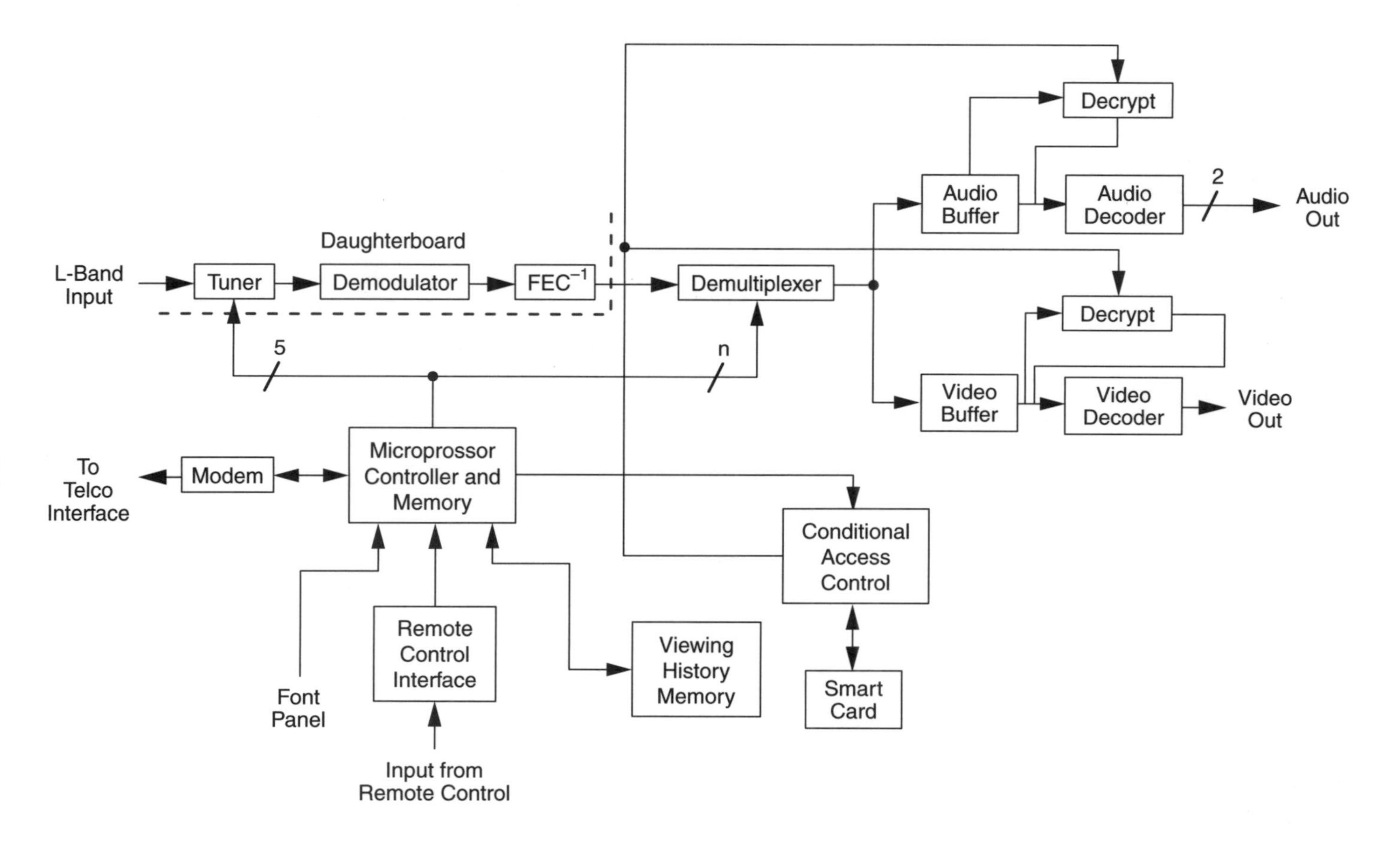

Figure 11.4 IDU Architecture

When requested by an over-the-air command, the MC connects to an access telephone number via modem. When this connection is made, the viewing history memory is transmitted to the Management Center. For valid subscribers, new keys are then sent to the IDU for the subsequent period.

Note that by having the viewing history remember the pay-per-view (PPV) services used and then downloading this to the Management Center for billing, true impulse PPV is obtained. All the consumer has to do is watch the desired service. There are no calls to make, no corresponding frequent busy signals, and so forth.

11.3 Electronic Program Guide

The Electronic Program Guide (EPG) differentiates DBS services from broadcast TV and C-band services, both of which require a separate paper pamphlet as a guide to the services offered. Also, the EPG allows IDU manufacturers to differentiate their services, because the graphical form of the EPG usually is not specified by the service provider.

Figure 11.5 is a screen image of the first-generation RCA-brand IRD Program Guide.[1] Note that the format is very similar to a newspaper listing with the service vertical (in increasing numerical order), and the time horizontal. Note that because of the large number of services and limited storage, only programs three hours in advance are shown. Published reports indicate that this is an area in which future improvements will be made.

11.4 Menu Selections

Figure 11.6 is a screen image of the main menu. Item 1 is the Program Guide previously discussed. Each of the menu items that follow permit the selection of other functions. This section is not intended to be a complete listing of all of the possible services. Instead, it illustrates how a modern system can simplify consumer operation.

If Options is selected from the main menu shown in Figure 11.6, Figure 11.7 results. The selection of setup results in Figure 11.8. Selecting item 1, Zip Code, results in Figure 11.9. At this point, enter the zip code where the antenna is located. Figure 11.10 shows this done for Carlsbad, California.

---

[1] This and all subsequent screen images in this chapter were obtained by connecting the S-Video output into a Macintosh 8500, then using the video viewer to capture the image.

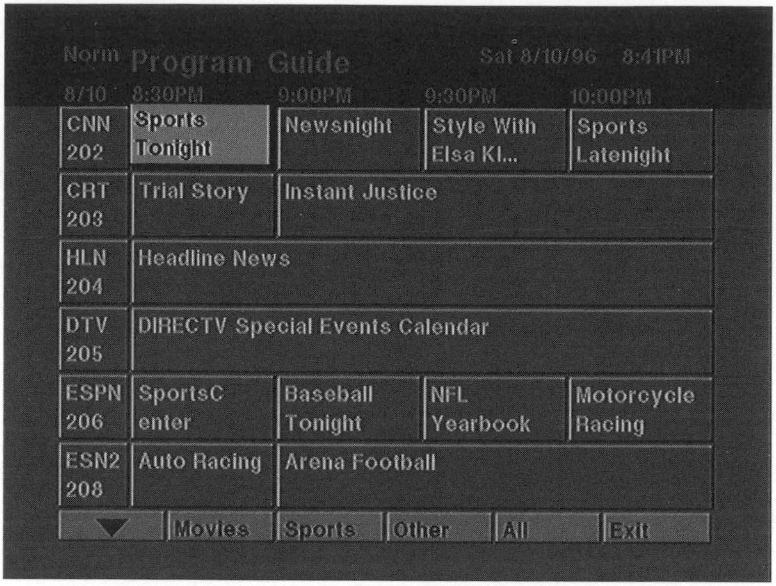

Figure 11.5 Program Guide Screen

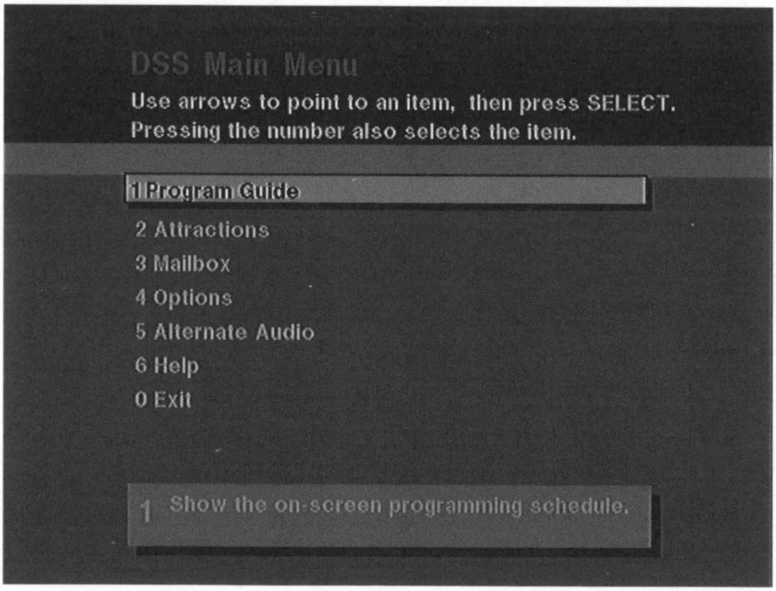

Figure 11.6 Main Menu

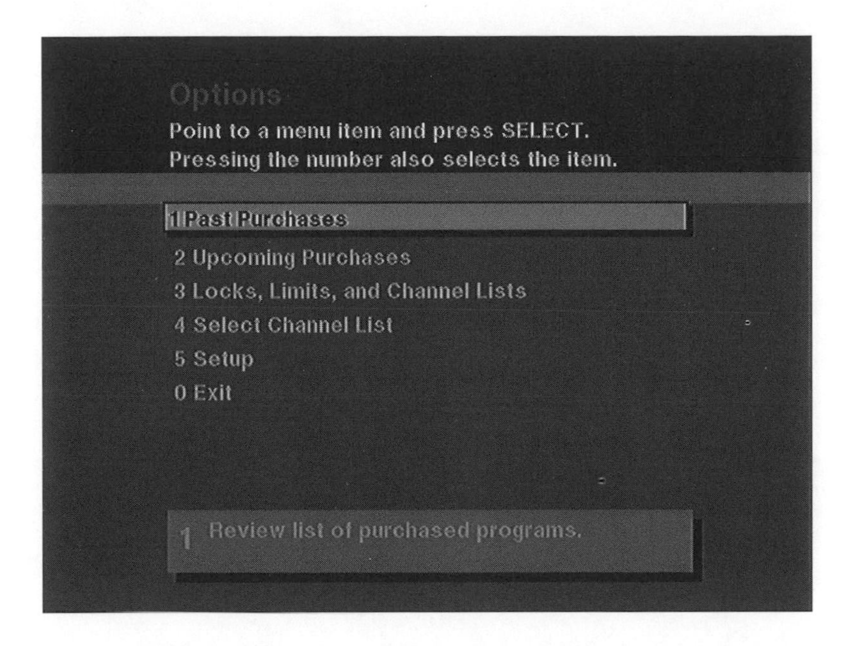

Figure 11.7 Options Menu

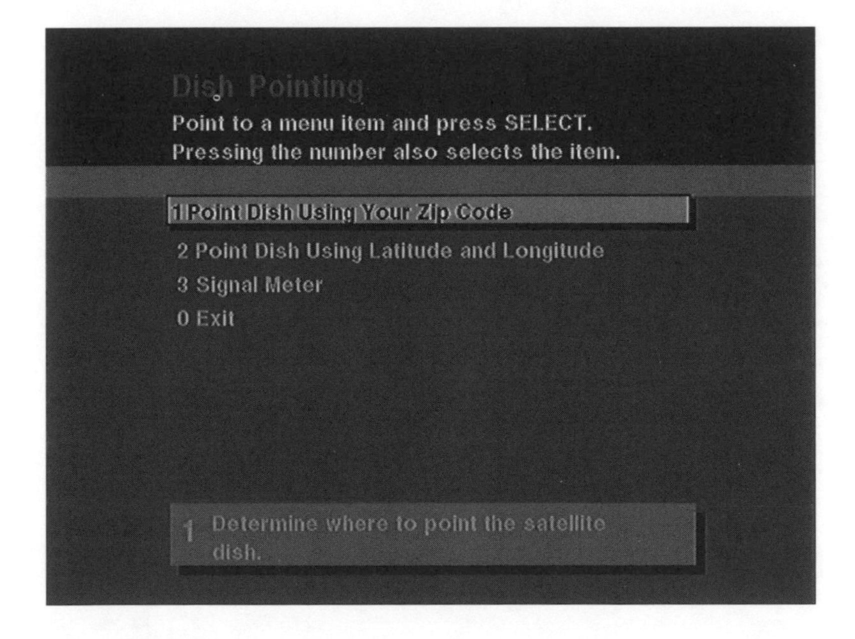

Figure 11.8 Dish Pointing

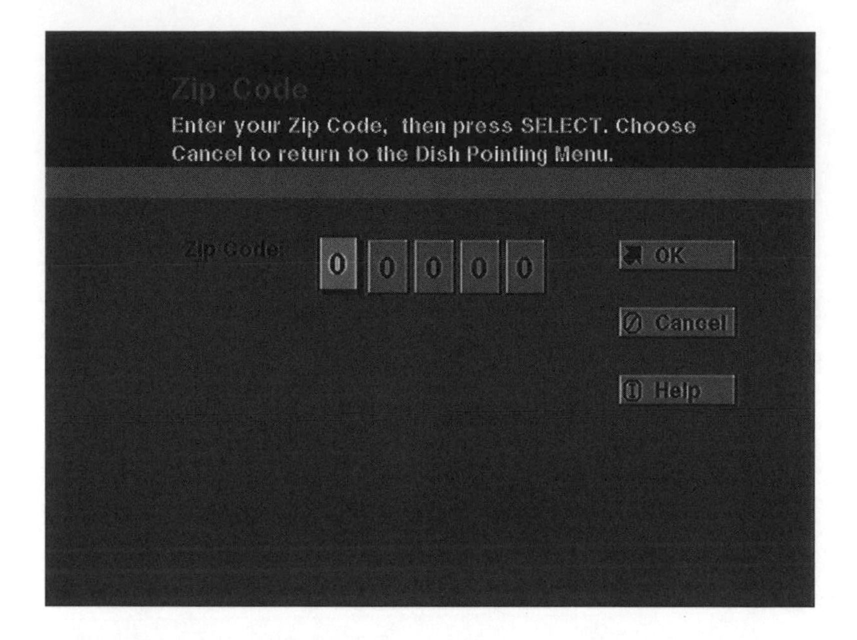

Figure 11.9 Zip Code

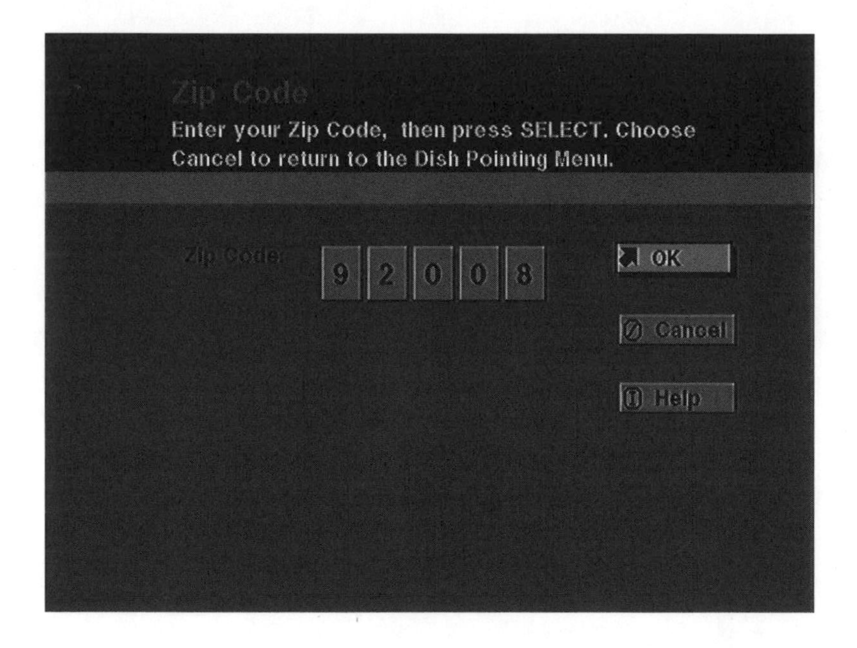

Figure 11.10 Zip Code Entered for Carlsbad, California

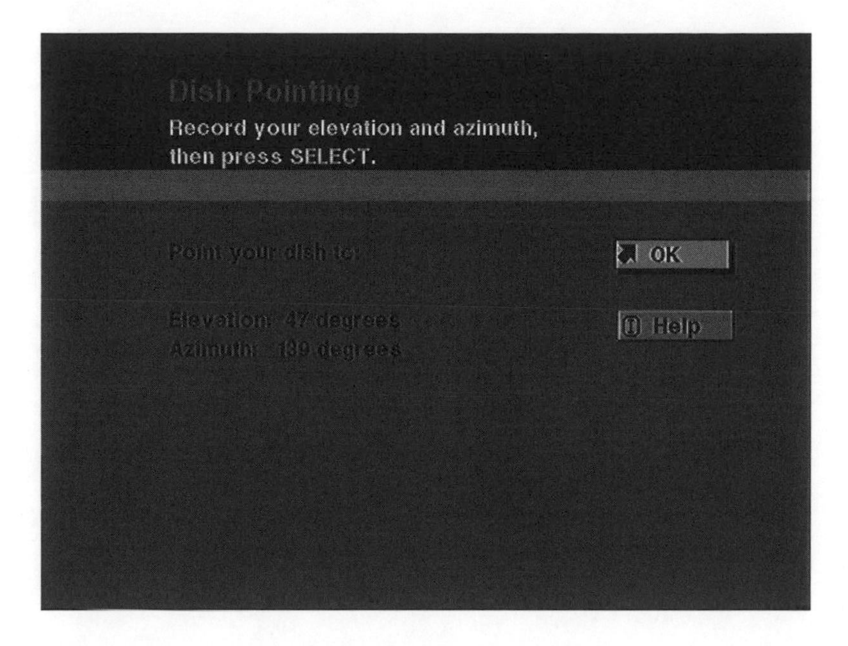

Figure 11.11 Elevation and Azimuth Angles for Carlsbad, California

The system responds with the correct elevation and azimuth angles, as shown in Figure 11.11. The antenna has registration marks that indicate the elevation. Thus, with the base of the antenna flat (measure it with a level), set the elevation to the correct angle. Select Signal Meter from the menu (Figure 11.12). Adjust the azimuth angle using an inexpensive compass until the signal meter creates a loud continuous tone. Finally, Figures 11.13 and 11.14 detail some of the control the customer has over content by showing the Locks and Limits menu and the Ratings Limit menu.

11.5 Multiple TV Sets

One of the most frequently asked questions in DBS is, I have more than one TV set. How do I get DBS for all of them? First, you have to have a decoder box for each TV set because there is only one decompress system in each of today's IRDs. Second, you have to use a dual LNB, with one LNB assigned to each of the two polarizations.

The dual LNB has two independent LNBs, one for each of the two polarizations. The first step is to force each of the two LNBs to select one of the polarizations by supplying 13 volts on the center conductor of the coax

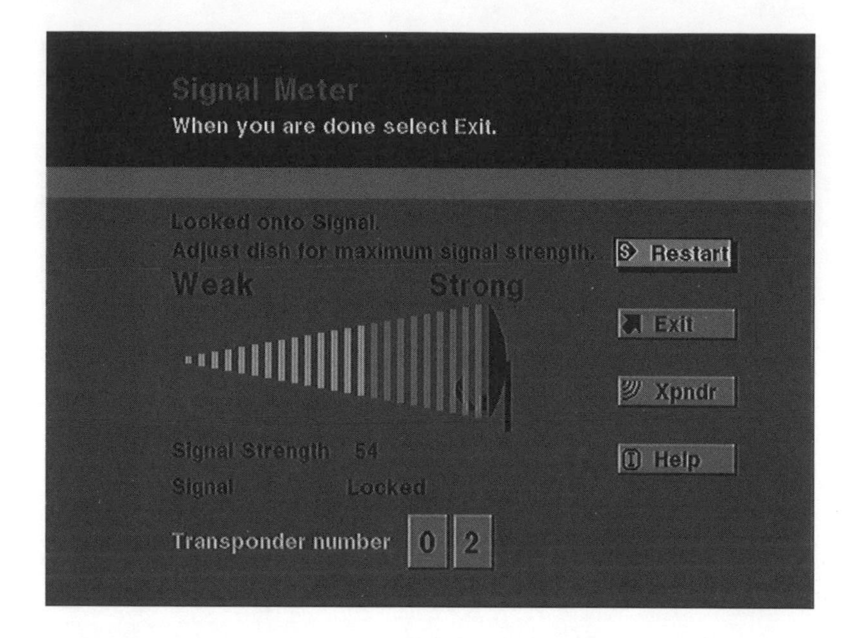

Figure 11.12 Signal Meter

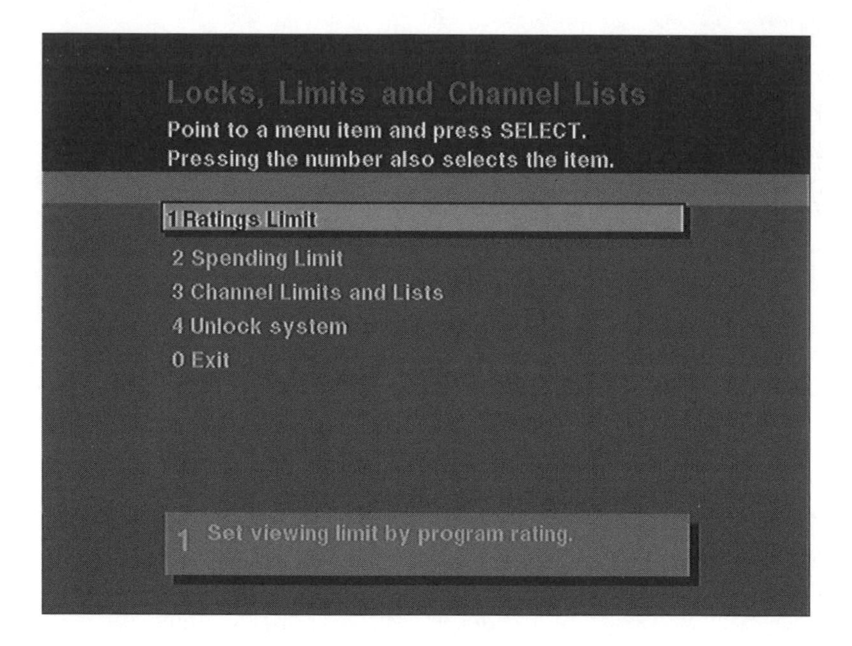

Figure 11.13 Locks and Limits

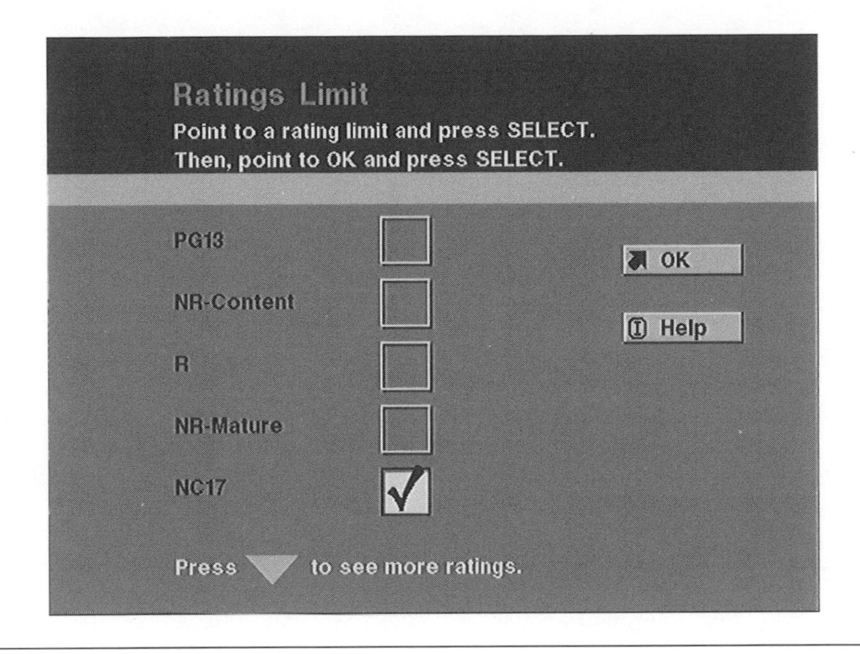

Figure 11.14 Rating Limits

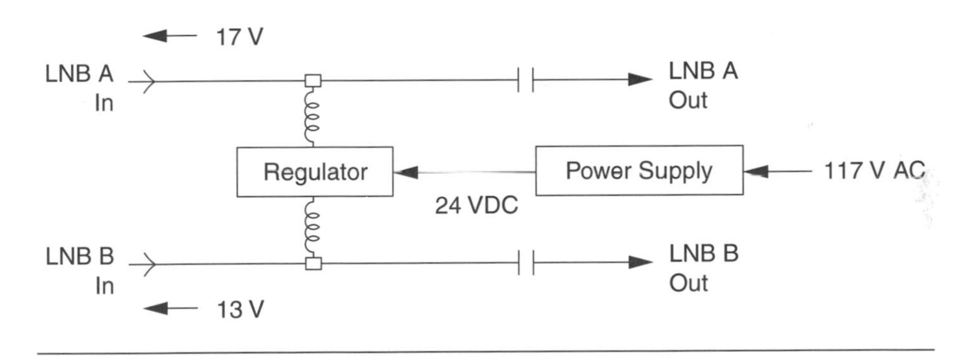

Figure 11.15 Dual LNB Power-Supply Injector

connection to the LNB and 17 volts to the other. This is done by the dual LNB power-supply injector shown in Figure 11.15. The capacitors shown in each leg keep the DC voltages from getting into the multiswitch shown in Figure 11.16.

The multiswitch receives the AC component from the LNB injector. It outputs this signal to the number of IRDs that are connected to it. Common multiswitch configurations are for 2, 4, 8, and 16 IRDs. The IRDs attach to the

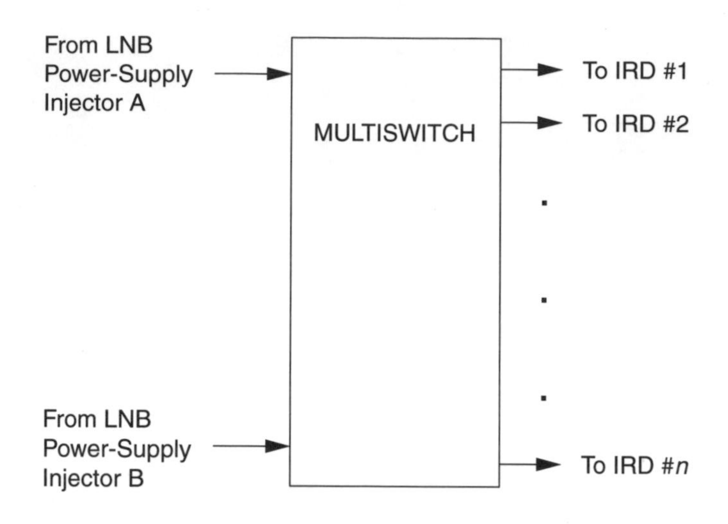

Figure 11.16 Multiswitch

multiswitch and provide a 13- or 17-volt signal to the multiswitch, which then provides one of the two polarizations to the IRD. Each IRD operates as if it had the entire antenna and LNB to itself, although they are really sharing them.

References

[Bursky95] Bursky, Dave, Single Chip Performs Both Audio and Video Decoding, *Electronic Design*, April 1995.

[Goodman96] Goodman, Robert L., *Digital Satellite Service*. New York: McGraw-Hill, 1996.

[Schweber95] Schweber, Bill, Direct Satellite Broadcast Brings Downlink Directly to Your Set-top Box, *EDN*, December 1995.

PART FIVE

The Future

CHAPTER TWELVE

Spaceway and the Global Broadcast Service

Most of this book has been about direct broadcast satellite (DBS) communications. This is a telecommunications broadcast medium that is called a *point-to-multipoint* transmission. In other words, a single uplink (Castle Rock, Colorado, for DIRECTV; Cheyenne, Wyoming, for EchoStar; and so on) transmits to millions and potentially tens of millions of receivers in the United States. However, there is no way for the subscriber to communicate back with the uplink.

Note that there is a communication capability in the IRD (see Chapter 11) for the report-back feature. However, this is controlled by DIRECTV and not the subscriber. Further, the communication is very asymmetric; the communication to the subscriber is wideband while the backlink is very narrowband.

A number of potential applications require two-way or duplex communication. For example, a video teleconference is not of much value if only one of the parties can be seen; thus, there is a need for wideband point-to-point communications. The Hughes Spaceway Satellite was conceived to fill these needs.

> *Anecdotal Note:* In the early 1990s when Spaceway began, Dr. Harold A. Rosen was a new grandfather. Video phones had small screens, terrible quality, and limited viewing angles. Harold believed there should be a capability for much higher quality so that people could dial the party they wanted to reach and be able to have an audio/video conversation. We designed the first implementation of this in my studio, and the project quickly became known as the Grandparent Project. As time went on, Hughes developed the business case, and so forth, Spaceway became the name, but it had its genesis in Harold's Grandparent Project [Fitzpatrick95].

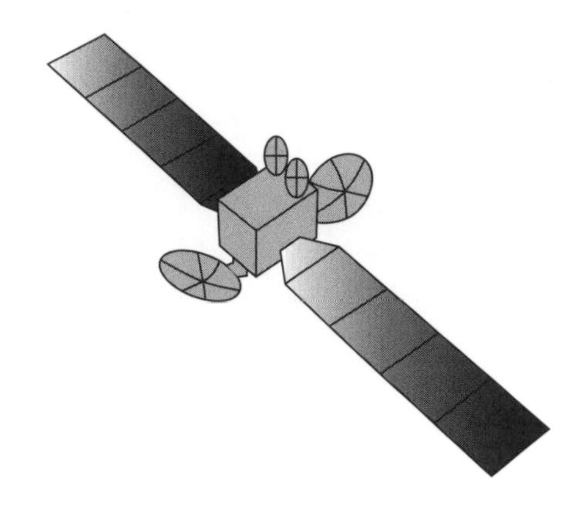

Figure 12.1 Spaceway Satellite

12.1 Spaceway Overview

Spaceway is designed to provide the point-to-point capability previously mentioned. It is planned to operate in the K_a-Band, which generally means that the uplink is at 30 GHz and the downlink is at 20 GHz [Careless96]. Hughes's studies have shown that a 26-inch-diameter antenna can provide reliable transceiver capabilities. The Spaceway satellite will be an enhanced version of the body-stabilized satellite used for DBS (see Figure 12.1). Spaceway will employ multiple spot beams, as shown schematically in Figure 12.2 for coverage of North America.

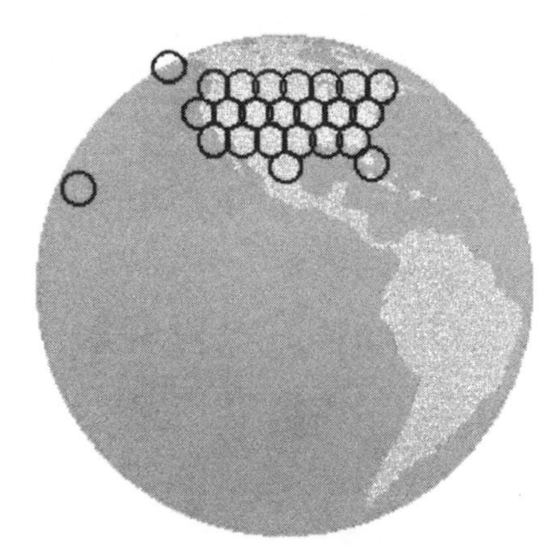

Figure 12.2 Spaceway Frequency Reuse

Each of the satellites will provide a total of 48 125-MHz spot beams (24 beams used on opposite circular polarizations) for uplink and downlink transmissions. In this way, each satellite effectively reuses the 500 MHz of spectrum assigned to it about 12 times. The original Hughes application to the FCC in December 1993 was for a two-satellite architecture that would be located at 101° W longitude and service the United States.

The key to the bandwidth-on-demand capability of the system is the onboard switching. Figure 12.3 is a block diagram of the satellite-switching subsystem. In it each of the 24 locations that receive spot beams is separated into its left-hand circularly polarized (LHCP) and right-hand circularly polarized (RHCP) components. Each of these components then is amplified by a low-noise amplifier (LNA), downconverted, and channelized. Next, the channelized outputs are demodulated and demultiplexed. At this point, the signals can be routed to the correct transmit spot beam.

Each of the downlink channels is QPSK modulated and amplified, then the two polarizations are combined to form the 24 downlink spot beams. The fact that the uplink signal is completely demodulated and that the forward error correction is removed means the signal is completely regenerated in the satellite. Thus, the uplink and downlink are completely decoupled. This is particularly advantageous in a duplex situation where the transmitters can be small terminals.

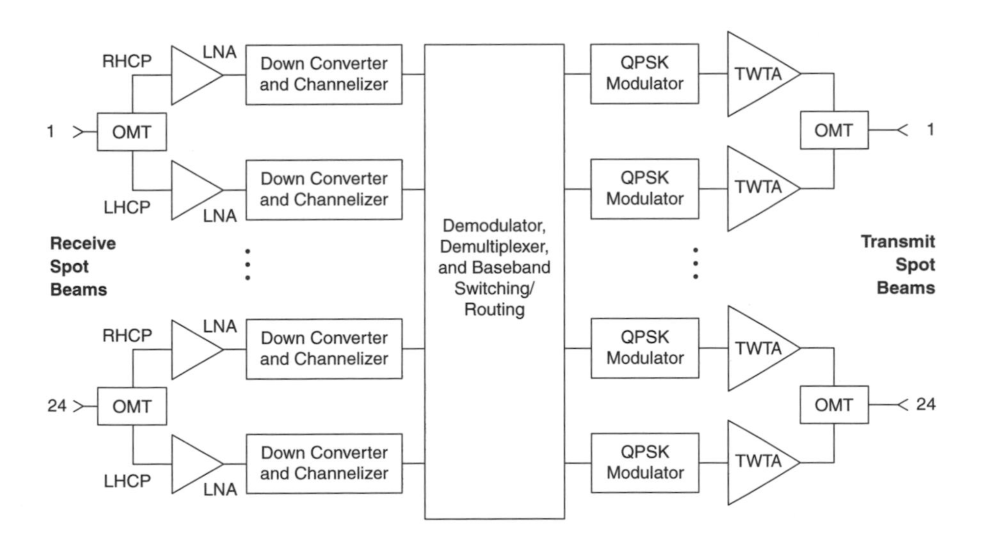

Figure 12.3 Onboard Communications System

12.2 Global Spaceway

As the Spaceway concept continued to evolve at Hughes, it became clear that the Spaceway should be a global system. Thus, on July 26, 1994, Hughes filed an amended application for a Global Spaceway system, which is to be implemented in two phases.

The first phase is a 9-satellite system, shown in Figure 12.4. As can be seen from the figure, North America is served by two satellites at 101° W longitude. The Pacific Rim is served by two satellites at 110° E longitude, Europe and Africa by two satellites at 25° E longitude, and South and Central America is served by two satellites at 50° W longitude. The ninth satellite at 175° E is called the Pacific Interconnect satellite and is to handle the large volume of traffic between Asia and the United States.

A study of Figure 12.4 shows that in addition to the ground beams, lines are drawn from each of the four pairs of satellites to each of the other three pairs. These represent intersatellite links (ISLs) that permit point-to-point communications when the two points are not within the footprint of a single

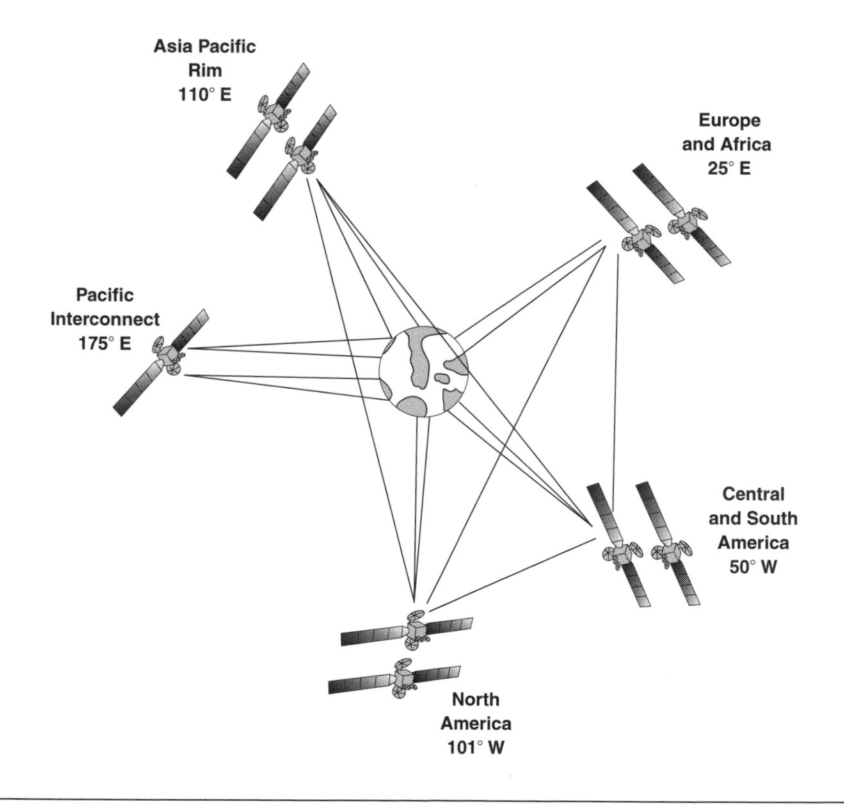

Figure 12.4 Spaceway Global Network—Phase 1 Orbital Locations

satellite, without an additional ground hop. Figures 12.5a, b, c, and d show the spot beam patterns for the first phase.

All the beams shown in Figures 12.5a–d are narrow-spot beams of 1 degree that have a 650-kilometer diameter at the Earth's surface. With 20-watt TWTAs, the EIRP at the spot-beam center is 59 dBw. In these narrow-spot beams, the G/T is 18.9 dB/K at the center of the spot beam and 13.9 dB/K at the edge of coverage.

a

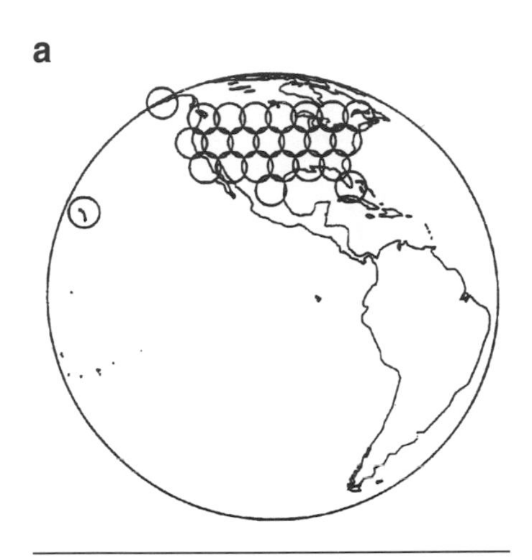

Figure 12.5a Phase 1, North America—Satellites 1 and 2

b

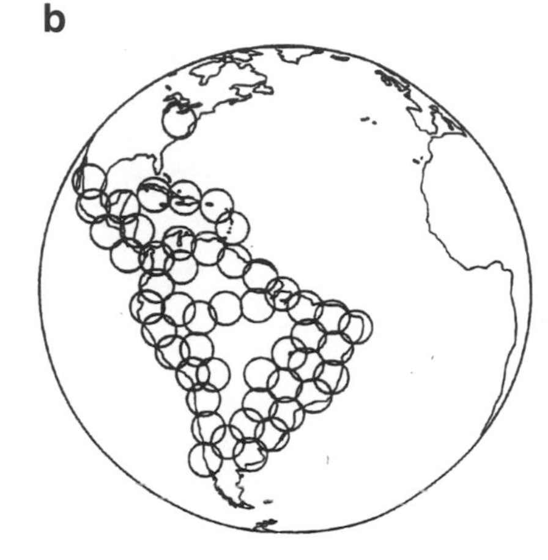

Figure 12.5b Phase 1, Central and South America—Satellites 1 and 2

c

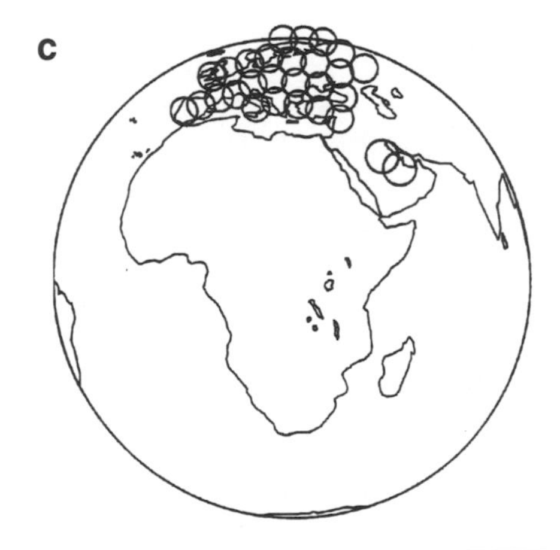

Figure 12.5c Phase 1, Europe and Africa—Satellites 1 and 2

d

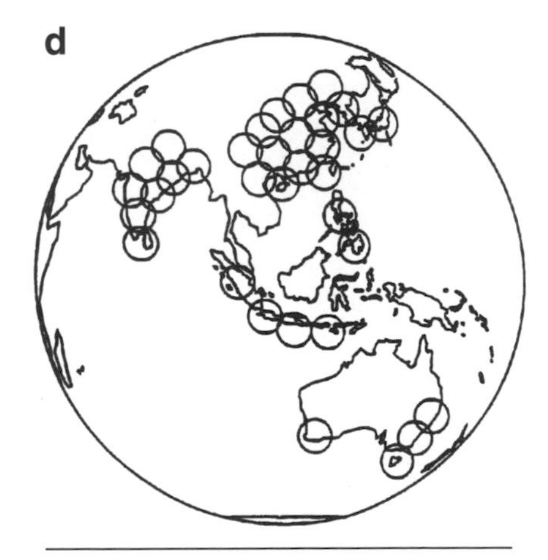

Figure 12.5d Phase 1, Asia Pacific Rim—Satellites 1 and 2

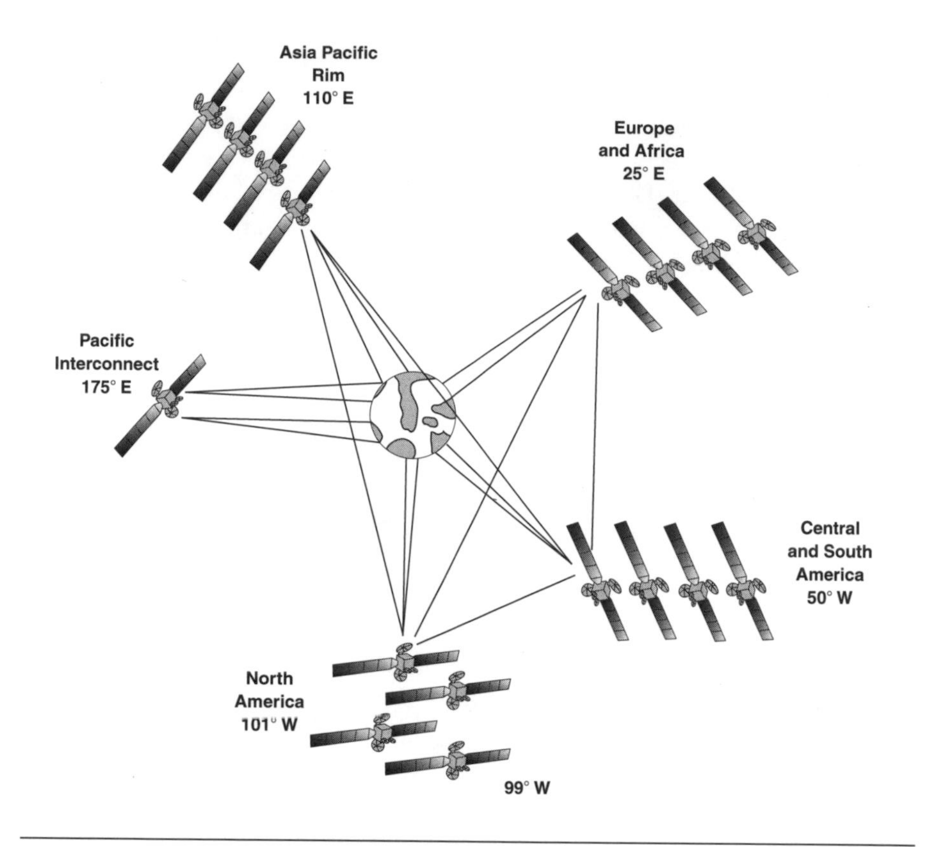

Figure 12.6 Global Spaceway Phase 2 Orbital Locations

The ISLs will be at a frequency of about 60 GHz. At these frequencies there is essentially 100 percent absorption by the Earth's atmosphere, so there is negligible effect on other services from the Spaceway.

Phase 2 of the Global Spaceway is to add eight satellites, essentially with four pairs of satellites joining the original four pairs. The lone exception is for North American coverage where the new pair is located at 99° W longitude rather than 101°. Figure 12.6 shows this final architecture.

Figures 12.7a, b, c, and d show the spot-beam patterns for the four sets of four satellites. A look at Figures 12.7a–d shows a mix of narrow-diameter spot beams and larger spot beams. In general, the larger beams are used to cover parts of the world with low population density. The larger spot beams are 3 degrees in beam width and have a 1,950-kilometer diameter at the Earth's surface.

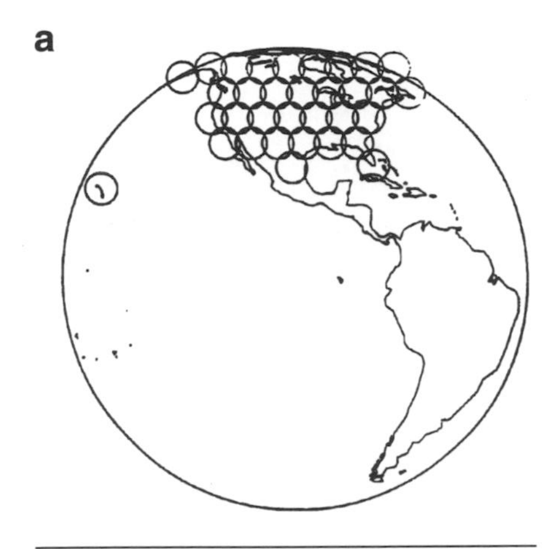

Figure 12.7a Phase 2, North America

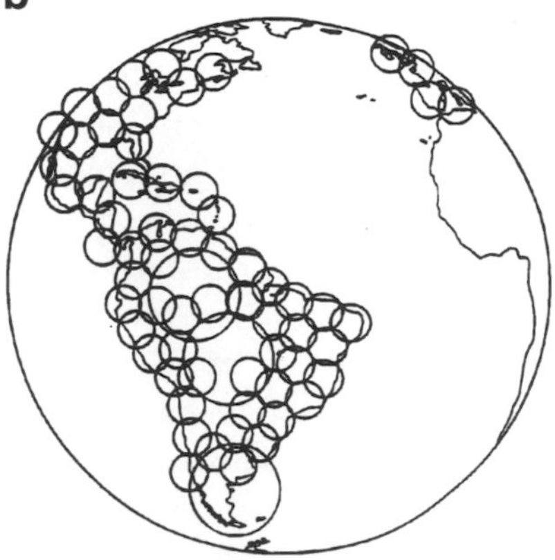

Figure 12.7b Phase 2, Central and South America

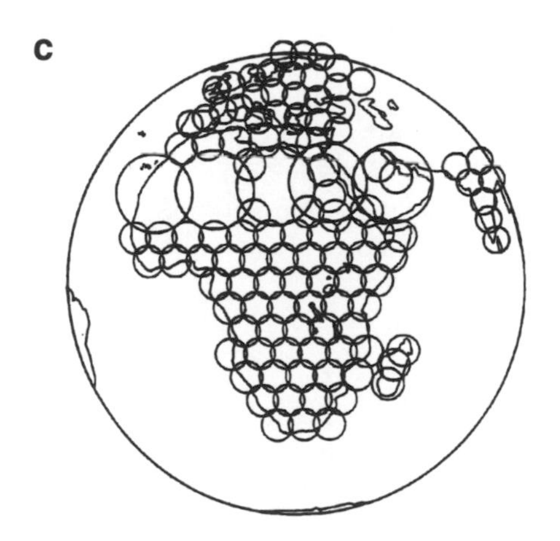

Figure 12.7c Phase 2, Europe and Africa

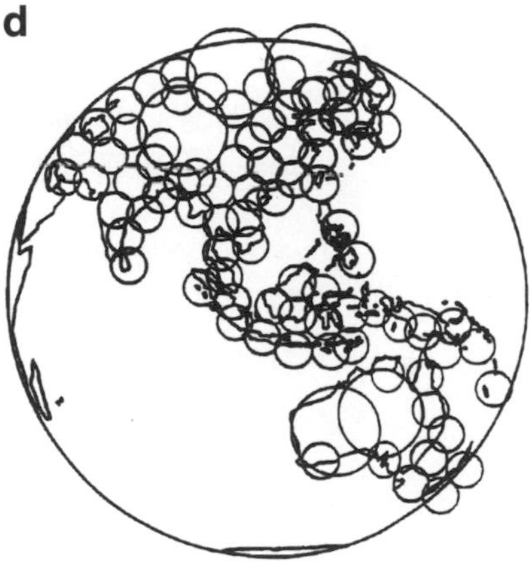

Figure 12.7d Phase 2, Asia Pacific Rim

The larger-diameter spot beams use a 60-watt TWTA and have an EIRP of 52 dBw at beam center. The G/T for the larger spot beam is 7.4 dB/K at beam center and 2.4 dB/K at the edge of coverage.

Consider the connectivity just indicated and the fact that the Spaceway will be able to transmit a high-resolution X-ray in less than 8 seconds that now takes 21 minutes, all with a terminal costing less than $1,000. Clearly, Spaceway is expected to be as revolutionary as DBS.

12.3 The Global Broadcast Service

DBS has major military implications. In modern warfare, information is power and the ability to communicate in hundreds of megabits per second is real power. The problem with commercially available DBS is that the EIRP is matched to locations where revenue-producing humans can receive the services (not fish in the oceans, for example).

The solution to this problem is movable-spot beams. In the Global Broadcast Service (GBS) Phase 2, under contract with the U.S. Navy, Hughes is modifying the last three satellites of the UHF follow-on satellite series to have a DBS capability with movable-spot beams. Each satellite will have four transponders and three movable spot beams: two with a 500-nautical-mile diameter and one with a 2,000-nautical-mile diameter.

Phase 2 of the GBS is to be operational during 1999 and will give the U.S. military an unprecedented ability to communicate worldwide.

References

[Careless96] Careless, James, K_a-Band Satellites: A Guide to the Proposed U.S. Systems. Via Satellite, February 1996.

[Fitzpatrick95] Fitzpatrick, Edward, Spaceway: Providing Affordable and Versatile Telecommunication Solutions, *Pacific Telecommunications Review* 17(1), September 1995.

CHAPTER THIRTEEN

Intelligent Compression: MPEG 4

In Chapters 8 and 9 we discussed video and audio compression as they are utilized in today's DBS Systems without a discussion of how such techniques could be improved in the future. In this chapter, we discuss the future directions of compression.

First, consider video. In the 1930s, when television was being developed, the only display device available was the Cathode Ray Tube (CRT). The drive electronics for x and y deflection created lines of video, as shown in Figure 8.1. Later, when digital video began to emerge, it was natural to sample the horizontal lines to form a rectangular array of numbers representing each frame of video. Video, thus, was represented as a temporal sequence of these rectangular arrays of numbers.

13.1 Raster and Waveform Emulation versus Symbolic Compression

All the mainstream video compression approaches attempt to compress the digital video by reducing the redundancies, both spatially and temporally, in such a way as to minimize the objectionability to the human vision system. A good name for this type of video compression is *raster video compression.* Note that there is no attempt to adapt the technique to the content of the video.

This approach has the advantage of being able to compress whatever video is presented to it. On the other hand, in the general case, it comes nowhere near approaching the information theoretic minimum bit rate or storage that could be achieved.

13.1.1 Example—Video

Suppose we have an unmanned aerial vehicle (UAV) with an infrared (IR) camera in the nose that is to be sent over a battlefield to search for enemy tanks. The results are to be transmitted back to a command post. It is known that there are three types of tanks that the camera may see.

It is a given that the IR camera is monochrome, has a pixel intensity dynamic range of 12 bits, a raster size of 480-by-640 pixels, and generates 15 frames per second. Thus, the uncompressed bit rate is 55.3 Mbps. It is probable that raster compression can achieve 25:1 or so, requiring a bit rate of more than 2 Mbps. Alternatively, electronics that could examine each frame of video from the camera could be placed in the UAV to make one of the four decisions shown in Table 13.1.

If the code for each condition were transmitted each frame time, only 30 bits per second would be required. Even if the necessary collateral data was transmitted, only several hundred bits per second would be required, a compression ratio of 270,000 rather than the 25:1 of conventional compression techniques. This huge increase in compression is made possible by the symbolic representation (the code) of the images of the tanks.

13.1.2 Example—Speech

Digital speech was designed by the telephone companies as 8 bits per sample times 8,000 samples per second or 64,000 bits per second. Hundreds of man years of research and development have reduced this to a typical number between 4,800 and 9,600 bits per second for compressed speech. The new federal standard in the United States—2,400 bits per second—is pretty good but still has some impairments. Virtually all these techniques involve a direct representation of the speech waveform.

As an alternative, consider the voice coder (*vocoder*) shown in Figure 13.1(a) and 13.1(b). Referring to Figure 13.1(a), the analog voice waveform is

Table 13.1 UAV Decisions

Condition	Code
No tanks	00
Tank type 1	01
Tank type 2	10
Tank type 3	11

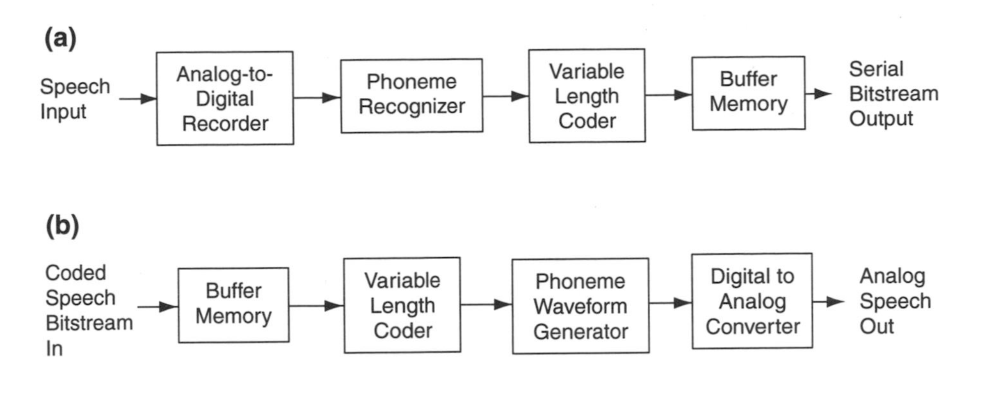

Figure 13.1 Symbolic Speech Encoder (a) and Decoder (b)

first converted to digital form. This digitized speech signal is broken into its constituent building-block sounds (called *phonemes*), and each phoneme is recognized. Next, a variable length code for that phoneme is looked up in the variable length codebook and placed in a buffer memory. The output of the buffer memory is a serial bitstream.

In the decoder the serial bitstream is first decoded by the variable length decoder. The output of the variable length decoder selects a phoneme waveform from the phoneme waveform generator that feeds the digital-to-analog converter. The output of the digital-to-analog converter is the analog speech waveform.

Although there are only about 100 phonemes in the international phoneme alphabet, a representation with 8 bits, or 256 possibilities, provides a rich representation space. Utilization of a variable length code reduces this to about an average of 4 bits per phoneme. Because the average phoneme is about 100 milliseconds, the bit rate of the symbolic vocoder of Figure 13.1 is 40 bits per second, or 1/60 of the new federal standard and 1/1,700 of the uncompressed bit rate.

Regardless of who speaks into the encoder of Figure 13.1(a), the sound from the decoder will be the same. This is because the output of the decoder is determined by the waveforms stored in the phoneme waveform generator. Thus, the vocoder of Figure 13.1(b) is not suitable for general telephony use. However, it could be valuable in special situations such as instructions for repairmen, and so on. Perhaps more important, by adding personalization layers to the vocoder (see following section), it can be made into a vocoder suitable for general telephony with a much lower bit rate than available today.

These two examples show that dramatic increases in coding efficiency are available with certain types of source materials. In addition to coding efficiency, many other functionalities are important for providing general digital audio/video (AV) services, which are often referred to under the general heading of *multimedia.*

13.2 MPEG 4: The First Steps Toward Symbolic Compression

13.2.1 Background

MPEG 1 and MPEG 2 were developed under enormous schedule pressure to meet the needs of the digital storage medium (DSM) and broadcast industries for compression standards. However, even as we were developing the MPEG 2 standard, it was clear to a number of us that there were applications for which there were no suitable standards.

Thus, during the MPEG meeting in Angra Dos Reis, Brazil, in July of 1992, a 10 PM meeting was called to discuss interest in "low bit rate coding." Perhaps all of us were astounded when 40 MPEG experts showed up after having already worked 13 hours on MPEG 2.

It was clear to me then and still is that "low bit rate coding" was a terrible name for the effort because what "low bit rate" means is in the eye of the beholder. To many it meant video telephones over plain old telephone system lines (POTS) or low bandwidth wireless networks; in other words, rates of under 64 Kbps. To broadcasters it meant TV programs at 384 Kbps that took 4 Mbps with MPEG 2, and for those interested in HDTV, it meant HDTV at T1 rates (1.544 Mbps) rather than the 18 Mbps required using MPEG 2. More about this difference in vision for MPEG 4 later.

There was sufficient interest in the effort, so it made its way through the ISO approval hierarchy and became a work item in 1994. When MPEG 4 becomes an international standard in 1999, it will be ISO 14,496.

13.2.2 The Enabling Technology for Multimedia

The following definition of multimedia was created by Dr. Leonardo Chiarliogne, the MPEG Convenor, and is the best I have seen (Private communication, 1995). *Multimedia communication* is the possibility of communicating audio/visual information that

1. Is natural, synthetic, or both
2. Is real time and nonreal time

3. Supports different functionalities responding to user's needs
4. Flows to and from different sources simultaneously
5. Does not require the user to bother with the specifics of the communication channel, but uses a technology that is aware of it
6. Gives users the possibility to interact with the different information elements
7. Lets users present the result of their interaction with content in the way that suits their needs

MPEG 4 is a current MPEG project that is being developed to provide enabling technology for the seven items here. Started in July 1993, it reached Working Draft level in November 1996, Committee Draft level in November 1997, and will reach International Standard level in 1999.

13.3 MPEG 4 Overview

The basic concept of MPEG 4 is that each scene is comprised of a set of objects, or building blocks, that can be combined in some fashion to recreate the scene. The objects can be quite general; for example, the fixed background of a scene can be considered as an object. Primitive AVOs are audio/video objects that will not be further subdivided.

13.3.1 Representation of Primitive AVOs

Figure 13.2 (MPEG working document m1394) represents an audiovisual scene that is composed of several AVOs, organized in a prescribed fashion. We can define primitive AVOs, including (1) a two-dimensional (2D) fixed background, (2) the picture of a talking person (without the background), and (3) the voice associated with that person.

MPEG standardizes a number of such primitive AVOs, capable of representing both natural and synthetic content types, which can be either two- or three-dimensional. In addition to the AVOs mentioned before, and shown in Figure 13.2, MPEG 4 defines the coded representation of objects such as the following:

- Talking heads and associated text to be used at the receiver's end to synthesize the speech and animate the head
- Animated human bodies
- Subtitles of a scene containing text and graphics

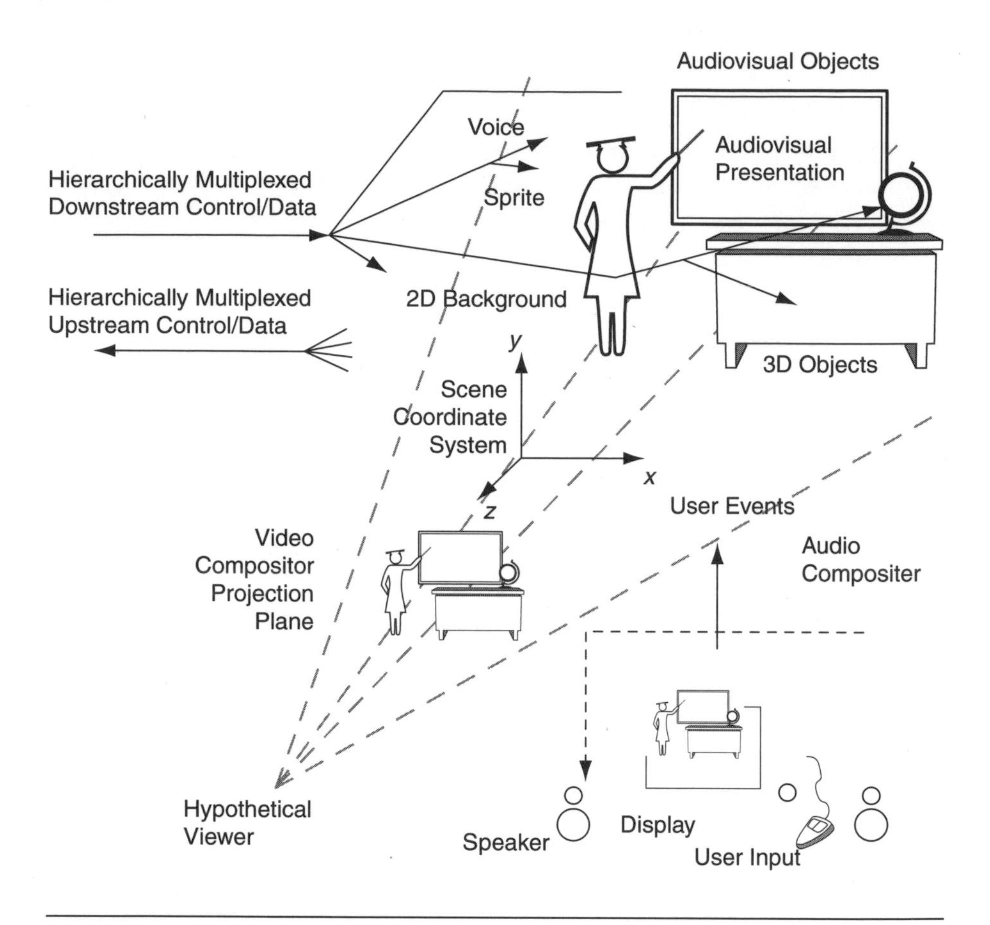

Figure 13.2 An Example of an MPEG 4 Audiovisual Scene

13.3.2 Composition of AVOs

The audiovisual scene of Figure 13.2 also contains compound AVOs that group arbitrary AVOs together. An example: The visual object corresponding to the talking person and the corresponding voice are tied together to form a new compound AVO—the audiovisual person.

Such grouping allows authors to construct complex scenes, and enables consumers to manipulate meaningful (sets of) objects. To use a modern metaphor, the primitive AVOs can be considered Lego bricks—the creators of content into meaningful content.

More generally, MPEG 4 provides a standardized way to compose a scene, for example allowing the following:

- Place AVOs anywhere in a given coordinate system
- Group primitive AVOs in order to form compound AVOs
- Apply streamed data to AVOs to modify their attributes (e.g., moving texture belonging to an object; animating a moving head by sending animation parameters)
- Change the users' viewing or listening positions anywhere in the scene

13.3.3 Multiplex and Synchronization of AVOs

In Chapter 7, MPEG 2 Systems, it was shown how packetized elementary streams (PES) were multiplexed together to form synchronized audio/video bitstreams. In a similar fashion, in MPEG 4, each AVO generates a PES that is then multiplexed with other PESs to create the overall scene.

13.3.4 Interaction with AVOs

Normally, the user observes a scene as it was composed by the scene's author. In MPEG 4, however, the user may also interact with the scene. Things a user might be able to do include:

- Changing the viewing/hearing point of the scene
- Rearranging objects in the scene to different positions, or deleting objects from a scene
- Triggering more complex kinds of behavior; for example, a virtual phone rings, the user answers, and a communication link is established

The content creator usually can specify the nature and degree of changes that can be made to the content.

13.4 Technical Description of the MPEG 4 Standard

The network layer delivers multiplexed bitstreams to elementary stream demultiplexers which recreate the original Elementary Streams, as shown in Figure 13.3 (m1394). The PESs are parsed and passed to the appropriate decoders. Decoding recovers the data of an AVO from its encoded format and performs the necessary operations to reconstruct the original AVO. The reconstructed AVO is made available to the composition layer for possible use during scene rendering.

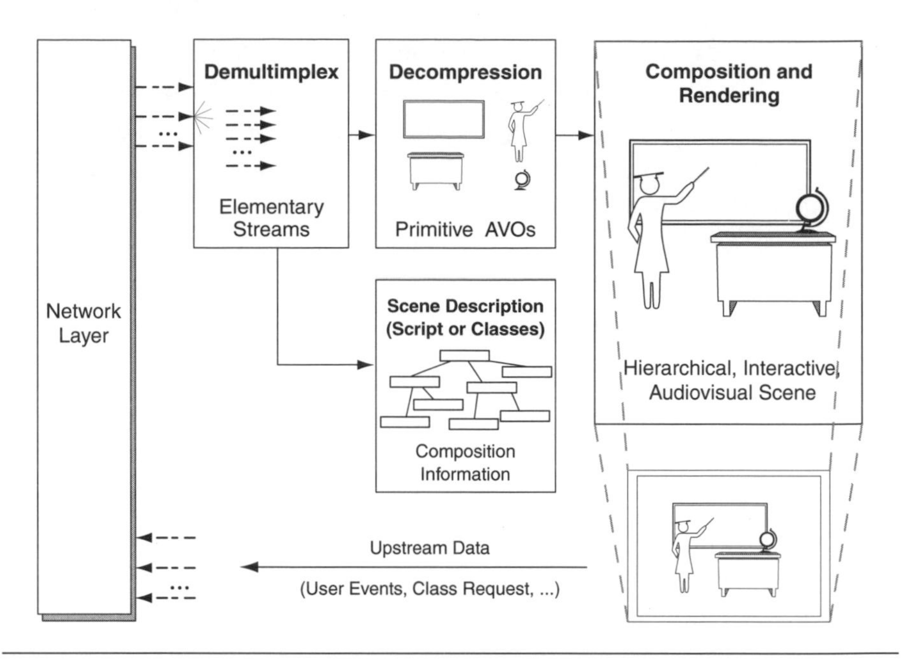

Figure 13.3 Major Components of an MPEG 4 Terminal (Receiver Side)

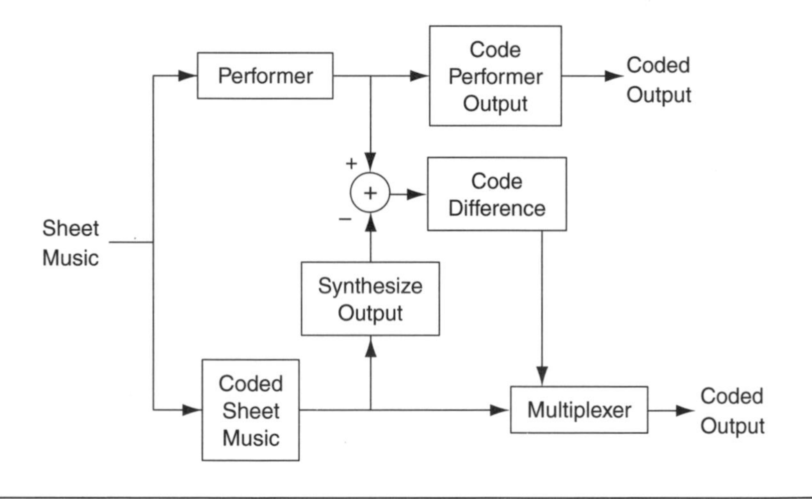

Figure 13.4 Audio Philosophy

Author's Note: MPEG 4 uses the language "natural" audio and "structured" audio. I believe this is an artificial distinction. The reason is shown in Figure 13.4. On the top half, a performer works from sheet music. The performer's output is then coded.

On the bottom half, the sheet music is coded, it is synthesized and compared with the performer's output. The difference is then coded and multiplexed with the coded sheet music to form the final output. When the two outputs are decoded, exactly the same output will ensue! However, the bottom approach generally will use significantly fewer bits.

Still, the MPEG Audio committee continues to use the separate designations—natural and structured.

13.4.1 Coding of Audio Objects

MPEG 4 coding of audio objects provides tools for representing natural sounds, such as speech and music, and for synthesizing sounds based on structured descriptions. The representations provide compression and other functionalities. The representation for synthesized sound can be formed by text or instrument descriptions and by coding parameters to provide effects, such as reverberation and spatialization.

MPEG 4 Audio provides different coding techniques for different audio requirements. For the first time in the history of MPEG, MPEG 4 has a specialized coding tool for speech. A parametric coder operating in the 2-Kbps to 6-Kbps range gives near toll-quality speech compression.

For general audio, Code Excited Linear Predictive (CELP) coding tools are available for bit rates between 6 Kbps and 24 Kbps. Finally, for multichannel audio, the Advanced Audio Coding (AAC) technique provides five channels plus low-frequency channel audio for high-quality applications.

Additional functionalities are realized both within individual coders and by means of other tools around the coders. An example of a functionality within an individual coder is pitch change within the parametric coder.

Synthesized Sound

Decoders for generating sound based on structured inputs are also available. Text input is converted to speech in the text-to-speech (TTS) decoder, while more general sounds, including music, are synthesized in accordance with a score description; this may include MIDI as the musical analog of text in TTS synthesis.

Text to Speech

Hybrid scalable TTS allows a text, or a text with parametric inputs, to generate intelligible synthetic speech. It includes the following functionalities:

- Facial animation control with Lip Shape Patterns or with phoneme information
- Trick mode functionality: pause, resume, jump forward or backward

- International language support for text
- International symbol support for phonemes

Score-Driven Synthesis

A script or score is a time-sequenced set of commands that invokes various instruments at specific times to contribute their output to an overall music performance or generation of sound effects. MIDI, an established protocol for transmitting score information, is a musical analog of the text input in the TTS synthesizer just described. The score description is more general than MIDI, which is a specific example because the score can include additional control information to allow the composer finer control over the final synthesized sound.

Effects

Structured audio/effects provide audio data so that an output data stream can be manipulated for special effects. The effects are essentially specialized "instrument" descriptions serving the function of effects processors on the input streams. The effects can include reverberators, spatializers, mixers, limiters, dynamic range control, filters, flanging, chorus, or any hybrid of these effects.

The audio sources can be natural sounds, perhaps emanating from an audio channel decoder or stored wave-table, thus enabling synthetic and natural sources to be merged with complete timing accuracy before being composited with visual objects at the system layer.

13.4.2 Coding of Visual Objects

Visual objects can be of either natural or synthetic origin. Objects of natural origin are described here.

Natural Textures, Images, and Video

The atomic units of image and video content are called *video objects* (VOs). An example of a VO is a talking person (without background) that can then be joined with other AVOs (audiovisual objects) to create a scene. Conventional rectangular imagery is handled as a special case of such objects.

The visual part of the MPEG 4 standard provides a toolbox that contains tools and algorithms to bring solutions for multimedia requirements. The MPEG 4 image and video coding algorithms give an efficient representation of visual objects of arbitrary shape, supporting content-based functionalities. It will support most functionalities already provided by MPEG 1 and MPEG 2, including the provision to efficiently compress standard rectangular sized image sequences at varying levels of input formats; frame rates; bit rates; and various levels of spatial, temporal, and quality scalability.

Content-based functionalities support the separate encoding and decoding of content (i.e., VOs). An important advantage of the content-based coding approach taken by MPEG 4 is that the compression efficiency can be significantly improved for some video sequences by using appropriate and dedicated object-based motion prediction "tools" for each object in a scene. A number of motion prediction techniques can be used to allow efficient coding and flexible presentation of the objects. This permits MPEG 4 compression to approach the information theoretic limits of compression.

For the hybrid coding of natural as well as synthetic visual data (for example, for virtual presence or virtual environments), the content-based coding functionality allows mixing a number of VOs from different sources with synthetic objects, such as a virtual background.

The coding of conventional images and video is achieved similar to conventional MPEG 1 and 2 coding and involves motion prediction and/or compensation followed by texture coding. For the content-based functionalities, where the image sequence input may be of arbitrary shape and location, this approach is extended by also coding shape and transparency information. Shape may be represented either by a binary mask or an 8-bit transparency component, which allows the description of transparency if one VO is composed with other objects.

Scalable Coding of Video Objects

MPEG 4 supports the coding of images and video objects with spatial and temporal scalability, both with conventional rectangular as well as with arbitrary shape. *Scalability* refers to the ability to decode only a part of a bitstream and reconstruct images or image sequences.

Scene Description

In addition to providing support for coding individual objects, MPEG 4 provides tools to compose a set of such objects into a scene. The necessary composition information forms the scene description, which is coded and transmitted together with the AVOs.

An MPEG 4 scene follows a hierarchical structure that can be represented as a tree. Each node of the tree is an AVO, as illustrated in Figure 13.5 (note that this tree refers back to Figure 13.2). The tree structure is not necessarily static; the relationships can change in time and nodes can be added or deleted.

In the MPEG 4 model, audiovisual objects have both a spatial and a temporal extent. Temporally, all AVOs have a single dimension, time (T). Each

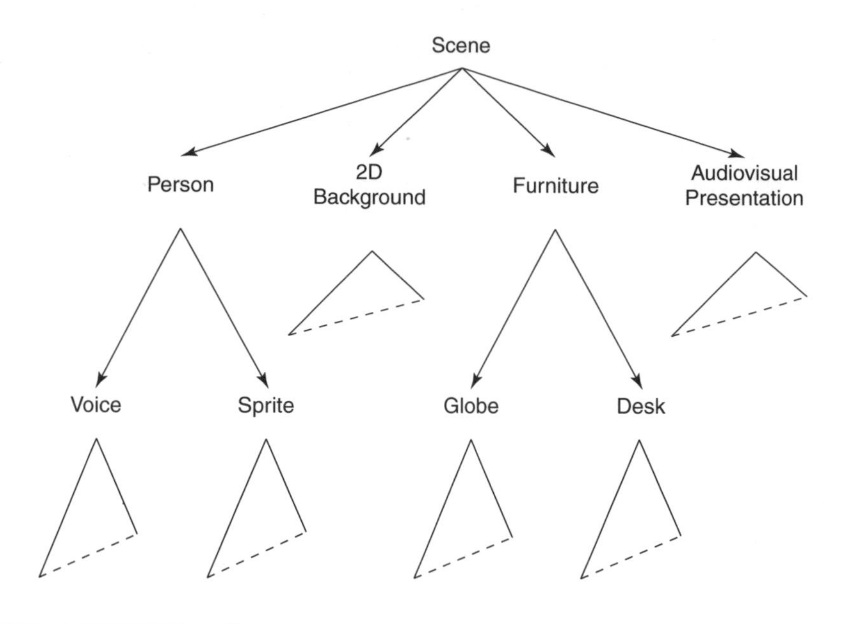

Figure 13.5 Logical Structure of the Scene

AVO has a local coordinate system. A *local coordinate system* for an object is a coordinate system in which the object has a fixed spatio-temporal location and scale. The local coordinate system serves as a handle for manipulating the AVO in space and time. Audiovisual objects are positioned in a scene by specifying a coordinate transformation from the object's local coordinate system into a global coordinate system defined by its parent AVO in the tree.

13.5 Conclusion

MPEG 4 will add a compression factor of 10 to 100 beyond that already achieved with MPEG 1 and MPEG 2. For certain content, the gains will be even greater.

MPEG 4 thinking is just in its infancy and figures to influence Audio and Video compression for at least the next 20 years.

APPENDICES

A Performance Degradation Because of Rain

B QPSK Modulation and Demodulation

C Algebraic Preliminaries

D BCH Code Details

E Cyclical Redundancy Code

F A5 Matrix

G Operators, Mnemonics, Abbreviations

APPENDIX A

Performance Degradation Because of Rain

For carrier frequencies above 10 GHz, rain will degrade performance. The following three different mechanisms contribute to this degradation.

1. Attenuation
2. An increase in sky noise temperature
3. Interference from adjacent channels when polarization is used for frequency reuse. This is caused by depolarization.

Each of these sources of degradation is calculated in the sections here, followed by a discussion of how each affects performance.

A.1 Rain Attenuation

Of particular interest for a DBS system is the percentage of time that certain values of attenuation are exceeded. The total attenuation is defined as

$$A_p = \alpha_p * L \tag{A.1}$$

where α_p equals the specific attenuation in dB per km, and L equals the effective path length in km.

The specific attenuation is defined as

$$\alpha_p = a * R_p^b \text{ dB/km} \tag{A.2}$$

where a and b depend on the frequency and the polarization. The parameters a and b for linear polarization are tabulated as a function of frequency.

Table A.1 Specific Attenuation Coefficients for Rain Attenuation Calculations

Frequency (GHz)	a_h	a_v	b_h	b_v
10	0.0101	0.00887	1.276	1.264
12	0.0188	0.0168	1.217	1.200
15	0.0367	0.0335	1.154	1.128
20	0.0751	0.0691	1.099	1.065
25	0.124	0.113	1.061	1.030
30	0.187	0.167	1.021	1.000

Note: Laws and Parsons drop size distribution; rain temperature equals 20°C.

The a and b parameters for circular polarization can then be calculated from their linear counterparts. Table A.1 gives the values for a and b as a function of frequency.

R_p is the rain rate in mm per hour and the effective path length L is equal to the rain slant length multiplied by a reduction factor r_p or $L = L_s * r_p$. The subscript p in each case refers to the fraction of a year that the parameter is exceeded.

These can all be combined to form the complete expression for rain attenuation:

$$A_p = a * R_p^b * L_s * r_p \tag{A.3}$$

For DBS the downlink frequency is approximately 13 Ghz, so from Table A.1, $a_h = .0188, a_v = .0168, b_h = 1.217, b_v = 1.2$. For circular polarization,

$$a_c = \frac{a_h + a_v}{2}$$

$$b_c = \frac{a_h b_h + a_v b_v}{2a_c}$$

The a and b parameters for circular polarization can be calculated as follows:

$$a_c = \frac{.0188 + .0168}{2} = .0178$$

$$b_c = \frac{(.0188) * (1.217) + (.0168) * (1.2)}{2 * (.0178)}$$

$$= \frac{.02288 + .02016}{2 * (.0178)} = 1.209$$

Thus,

$$\alpha_p = .0178 * R_p^{1.209}, \quad \text{for } f = 12\text{Ghz} \tag{A.4}$$

Taking into account the Earth's curvature,

$$L_s = \frac{2*(h_R - h_O)}{\left[\sin^2[\theta] + 2*\dfrac{(h_R - h_O)}{8{,}500}\right]^{1/2} + \sin[\theta]} \tag{A.5}$$

where

h_R = rain height

c = .6 for $\phi \leq 20°$

c = .6 + .02(ϕ – 20), 20° < $\phi \leq$ 40°

c = 1.0, ϕ > 40°

h_O = height above sea level of IRD (in km)

ϕ = IRD latitude

θ = elevation angle

In the continental United States, ϕ ranges from 24.5° to 49° and θ from 27° to 54°. Therefore, c ranges from .69 to 1.0. For c = .69,

$$\begin{aligned} h_R &= .69*[5.1 - 2.15*\log(1 + 10^{-.167})] \\ &= .69*\{5.1 - 2.15*\log[1 + .681]\} \\ &= .69*4.615 = 3.18 \text{ km} \end{aligned}$$

For the larger ϕ,

$$\begin{aligned} h_R &= \left[5.1 - 2.15*\log\left(1 + 10^{22/15}\right)\right] \\ &= 1.915 \text{ km} \end{aligned}$$

For the purposes of obtaining some numerical results, assume the calculation for close to sea level (h_O = .2 km) and use Equation (A.5).

$$L_s = \frac{2*(h_R - .2)}{\left[\sin^2\theta + 2*\dfrac{(h_R - .2)}{8{,}500}\right]^{1/2} + \sin\theta} \tag{A.6}$$

Table A.2 Slant Range Reduction Factor

	$\theta = 27°$	$\theta = 54°$
$r_{.001}$	.7485	.822
$r_{.01}$	.964	.977
$r_{.1}$	.982	.988

For the lower lattitudes, using $h_R = 3.18$ km and $\theta = 54°$, we have

$$L_s = \frac{2 * 2.98}{\left[.655 + 2 * \frac{5.96}{8,500}\right]^{1/2} + .809} = \frac{5.96}{1.619} = 3.68 \text{ km} \tag{A.7}$$

For the higher lattitudes, using $h_R = 1.915$ km and $\theta = 27°$, we have

$$L_s = \frac{2 * 1.715}{\left[.206 + 2 * \frac{3.43}{8,500}\right]^{1/2} + .454} = \frac{3.43}{.909} = 3.77 \text{ km} \tag{A.8}$$

Next, r_p can be determined from Table A.2. For $r_{.001}$,

$$r = \frac{10}{10 + L_G}$$

For $r_{.01}$,

$$r = \frac{90}{90 + 4L_G}$$

For $r_{.1}$,

$$r = \frac{180}{180 + L_G}$$

where $L_G = L_S * \cos\theta$. For $\theta = 54°$, $L_G = 3.68 * \cos[54°] = 2.16$ km, while for $\theta = 27°$, $L_G = 3.77 * \cos[27°] = 3.36$ km.

Finally, we can calculate the total attenuation for $\theta = 27°$ and $\theta = 54°$. For $\theta = 27°$,

$$A_p = .0178 * R_p^{1.209} * 3.77 * r_p$$

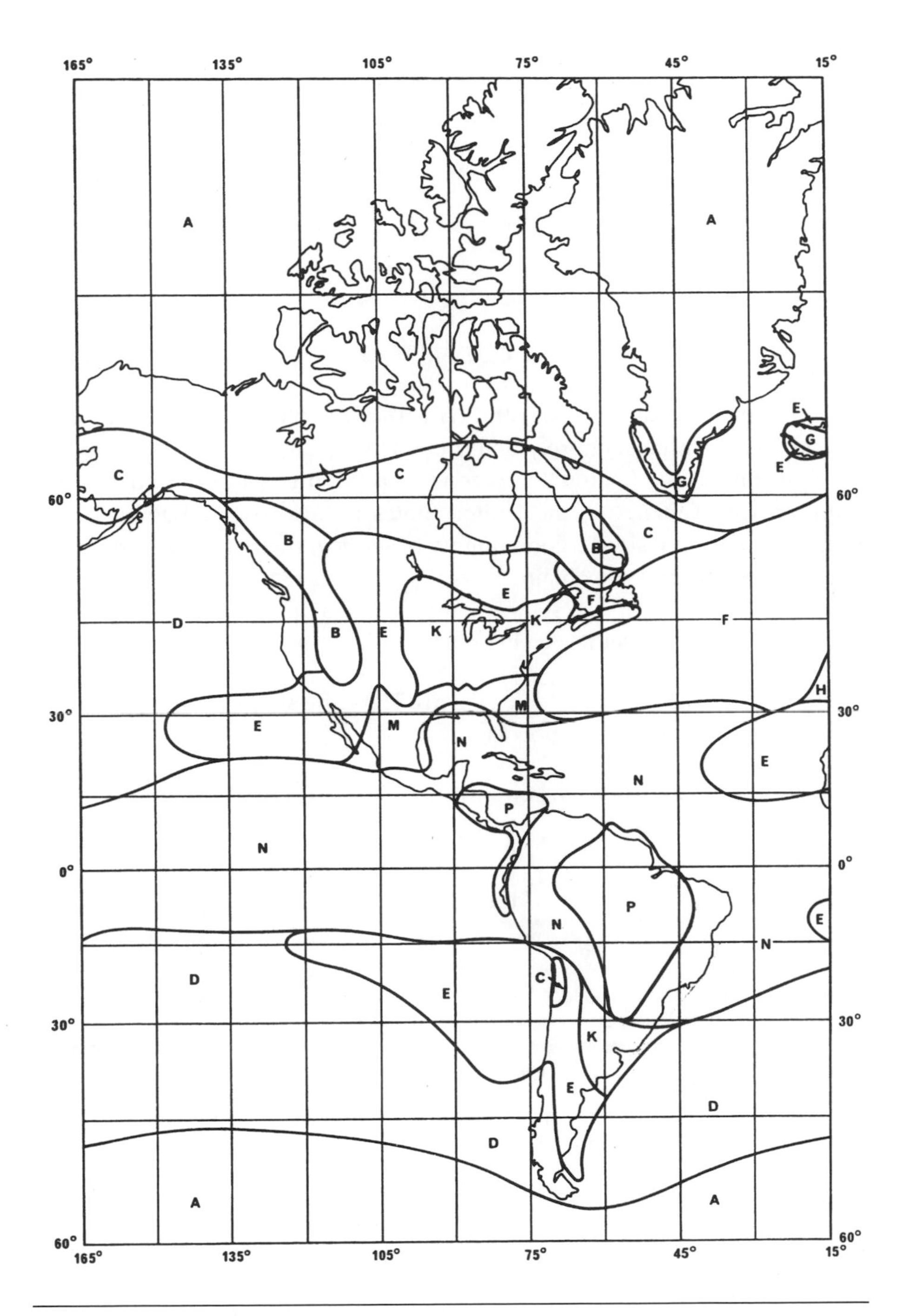

Figure A.1 Region 2 Rain Model

For $\theta = 54°$,

$$A_p = .0178 * R_p^{1.209} * 3.68 * r_p$$

For $\theta = 27°$,

$$A_p = .0671 * R_p^{1.206} * r_p \tag{A.9}$$

For $\theta = 54°$,

$$A_p = .0655 * R_p^{1.206} * r_p \tag{A.10}$$

Examination of Equations (A.9) and (A.10) shows that the attenuation in the United States is almost entirely determined by R_p.

Referring to the Region 2 rain zone map (Figure A.1) and rain rate distribution (see Table A.3), the United States ranges from a Region B to a Region M. If we consider a good quality of service, such as .1 percent outage per year (8.76 hours), we obtain

$$A_{.1}\text{(for Region B)} = .0671 * R_{.1}^{1.206} * .982 \tag{A.11}$$

Using $R_{.1} = 2$, we have $A_{.1} = .152$ dB. Thus, the attenuation is negligible. Next, however, consider Region M,

$$A_{.1}\text{(for Region M)} = .0655 * 22^{1.206} * .988$$

$$A_{.1}\text{(for Region M)} = 2.69 \text{ dB}$$

Thus, we see one of the reasons the margins are set high.

Table A.3 Rain Rate Distributions for the Rain Climate Zones of the CCIR Prediction Model

Percentage of Time (%)	Rain Rate Distribution Values (mm/h)													
	A	B	C	D	E	F	G	H	J	K	L	M	N	P
1.0	–	1	–	3	1	2	–	–	–	2	–	4	5	12
0.3	1	2	3	5	3	4	7	4	13	6	7	11	15	34
0.1	2	3	5	8	6	8	12	10	20	12	15	22	35	65
0.03	5	6	9	13	12	15	20	18	28	23	33	40	65	105
0.01	8	12	15	19	22	28	30	32	35	42	60	63	95	145
0.003	14	21	26	29	41	54	45	55	45	70	105	95	140	200
0.001	22	32	42	42	70	78	65	83	55	100	150	120	180	250

A.2 The Effect of Rain on System Noise Temperature

In addition to the attenuation of the signal power because of the rain, rain also increases the system noise temperature. This lowers G/T, further decreasing the carrier-to-noise (C/N). The increase in noise temperature is

$$\Delta T = T_r * \left(1 - \frac{1}{A_p}\right)$$

where T_r is normally taken to be 273°K. In the previous example,

$$A_{.1} = 2.69 \text{ dB} = 1.86$$

$$\therefore \Delta T = 273 * \left(1 - \frac{1}{1.86}\right) = 126°$$

Since G/T was 11.39 dB/°K, with $T_s = 125$ K,

$$\left(\frac{G}{T}\right)_{\text{rain}} = \frac{1,720}{251} = 8.36 \text{ dB/°K}$$

Thus, the rain has caused another 3 dB drop in the C/N.

A.3 Adjacent Channel Interference and Depolarization Effects Due to Rain

Adjacent channels are separated by 14.28 Mhz (see Figure 2.1). These channels are isolated by the fact that they have opposite polarization (right-hand circular or left-hand circular). With clear sky conditions, this isolation is 26 dB and does not significantly affect the C/N. However, rain causes a depolarization that creates adjacent channel interference.

Appendix B shows that QPSK can be considered to be the sum of two independent BPSK modulation paths, each operating at a bit rate equal to the QPSK symbol rate. Referring to Figure B.1, each path consists of the digital signal being passed through a transmit low-pass filter, multiplied by a sinusoid, and then output summation. The multiplication by the sinusoid translates the Power Spectral Density (PSD) in frequency but does not change its shape. Likewise, the summation does not change the shape of the Power Spectral Density. Thus, we can concentrate on the PSD of the digital signal and its modification by the transmit low-pass filter.

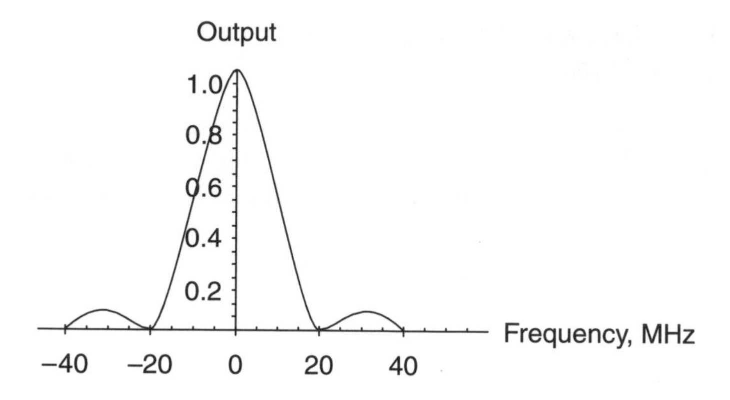

Figure A.2 Bipolar Power Spectral Density

First, it can be shown [Ha90, Appendix 9B][1] that the PSD for bipolar signaling is

$$s(f) = c * \left(\frac{\sin(\pi f T_s)}{\pi f T_s}\right)^2$$

where c is a constant, and T_s is the symbol period. For the major DBS systems, $T_s = 50 * 10^{-9}$ seconds. In this expression for $s(f)$, $f = 0$ corresponds to the carrier frequency. Figure A.2 is a plot of $s(f)$.

Because the transponder bandwidth is only 24 Mhz, and the transponder centers only 29 Mhz, it is clear that the output of the bipolar signal needs to be band-pass filtered. To see this, look at the value of the output at 14.5 MHz on either side of 0 in Figure A.2. The output hasn't even reached the first null. Thus, there is main-lobe energy and all the side-lobe energy in the adjacent transponders.

A raised cosine filter is defined by

$$V(f) = \begin{cases} 1, & f \le f_1 \\ .5 * \left[1 + \cos\left(\frac{\pi * (f - f_1)}{(B - f_1)}\right)\right], & f_1 \le f \le B \\ 0, & f \le B \end{cases}$$

[1] Ha, Tri. *Digital Satellite Communications.* New York: McGraw Hill, 1990.

The parameters f_1 and B are defined by a parameter called the roll-off factor $\rho, 0 \leq \rho \leq 1$, as

$$B = \frac{(1+\rho)}{2} * R_{symbol}$$

$$f_1 = \frac{(1-\rho)}{2} * R_{symbol}$$

For the major DBS systems, ρ is chosen to be .35. Because $R_{symbol} = 20 * 10^6$ symbols/sec,

$$B = \left(\frac{1+.35}{2}\right) * 20 * 10^6$$
$$= 13.5 * 10^6 \text{ sec}^{-1} = 13.5 \text{ MHz}$$

and

$$f_1 = \frac{.65}{2} * 20 * 10^6$$
$$= 6.5 * 10^6 \text{ sec}^{-1} = 6.5 \text{ MHz}$$

$$\therefore V(f) = \begin{cases} 1, & f \leq 6.5 \text{ MHz} \\ \left[1 + \cos\left(\frac{\pi * (f - 6.5 * 10^6)}{(13.5 - 6.5) * 10^6}\right)\right], & 6.5 \text{ MHz} \leq f \leq 13.5 \text{ MHz} \\ 0, & f \leq 13.5 \text{ MHz} \end{cases}$$

Figure A.3 shows $V(f)$. The Power Spectral Density of the filter output is

$$S_{out}(f) = S_{bipolar}(f) * |V(f)|^2$$

Therefore,

$$S_{out}(f) = \left(\frac{\sin(\pi f T_s)}{\pi f T_s}\right)^2, \quad 0 \leq f \leq 6.5 \text{ MHz}$$

$$= \left[.5 * \left(1 + \cos\left[\frac{(f - 6.5)}{7} * \pi\right]\right)\right]^2 \left[\frac{\sin(\pi * f * T_s)}{\pi * f * T_s}\right]^2,$$

for 6.5 MHz $< f <$ 13.5 MHz

$= 0, f >$ 13.5 MHz

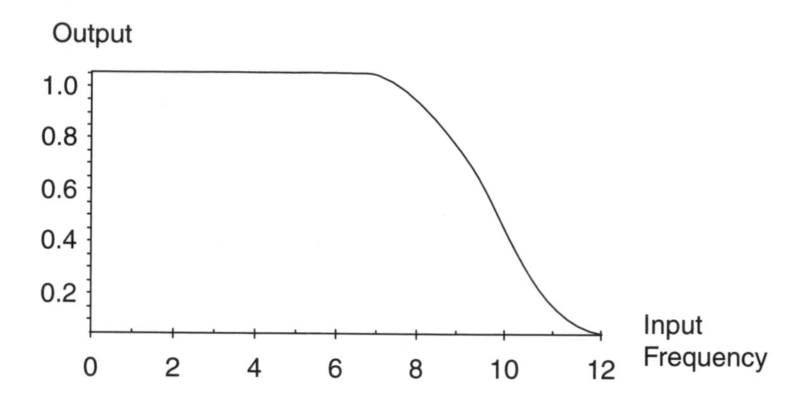

Figure A.3 Raised Cosine Filter

Figure A.4 shows this output PSD. It should be noted that $V(f)$ is a band-pass filter that is symmetric in negative frequency. The power in-band (in arbitrary units) is

$$\begin{aligned}
P &= \int_{-12}^{12} S_{\text{out}}(f) * df \\
&= 2 * \int_{0}^{12} S_{\text{out}}(f) * df \\
&= 2 * \left[\int_{0}^{6.5} S_{\text{out}}(f) * df + \int_{6.5}^{12} S_{\text{out}}(f) * df \right] \\
&= 2 * [5.80667 + 1.47978] \\
&= 14.5729
\end{aligned}$$

With this filtering, there is negligible adjacent channel interference from the same polarization. However, the adjacent channel of the opposite polarization will interfere when it is depolarized by rain.

Plotted on Figure A.4, it would be centered at 14.58 MHz and would be the curve of Figure A.4 translated to the right by 14.58 MHz. Its power, in the in-band (the interfering power), is

$$\begin{aligned}
I &= \int_{1.08}^{135} S_{\text{out},1458}(f) * df \\
&= \int_{1.08}^{8.08} S_{\text{out},1458}(f) * df + \int_{8.08}^{13.5} S_{\text{out},1458}(f) * df \\
&= .0192606 + 4.73011 = 4.74937
\end{aligned}$$

This is the power from the adjacent opposite polarization band. It is lowered by the cross-polarization discrimination before it becomes in-band interference.

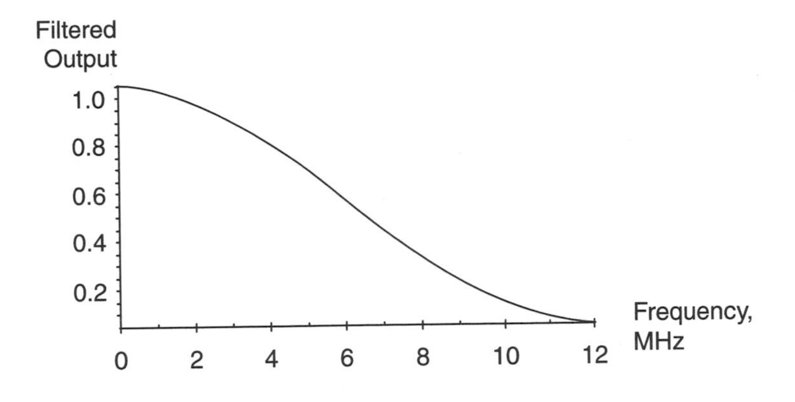

Figure A.4 Output Power Spectral Density

The cross-polarization discrimination, *XPD*, is an empirically determined value,

$$XPD = U - V * \log A_p$$

where

$$V = \begin{cases} 20, & \text{for } 8 \le f \le 15 \text{ GHz} \\ 23, & \text{for } 15 \le f \le 35 \text{ GHz} \end{cases}$$

and

$$U = 30 * \log f - 10 * \log[.5 - .4697 * \cos(4\tau)] - 40 * \log[\cos[\theta]]$$

where τ = tilt angle of polarization with respect to horizontal; $\tau = 45°$ for circular polarization, and θ = elevation angle. Using $\theta = 45°$,

$$U = 30 * \log[12.45 * 10^9] - 10 * \log[.5 - .4697 * (-1)] - 40 * \log[.707]$$
$$= 398.89 \approx 26 \text{ dB}$$

Thus, for DBS,

$$XPD = 26\text{–} 20 * \log A_p$$

For clear sky conditions, $A_p = 1$ and $XPD = 26$ dB. The effective adjacent cross-polarization band power is reduced by this amount to .0119, which is negligible.

However, when there is rain, $A_p > 1$ and the *XPD* is reduced. From the example earlier in this section, $A_{.1} = 1.86 = 2.69$ dB and

$$\begin{aligned} \textbf{XPD} &= 26 - 20 * \log 1.86 \\ &= 26 - 5.69 = 20.31 \text{ dB} \\ &= 100 \end{aligned}$$

Because there are interfering transponders on both sides, the total relative interfering power is 2(4.74937)/100 = .095. In this example then,

$$\frac{C}{I} = 153 = 21.85 \text{ dB}$$

A.4 Total Effects from Rain

We are now in a position to calculate the total effects of rain on the signal-to-noise:

$$\begin{aligned} \text{Attenuation, } A_{.1} &= 2.69 \text{ dB} \\ \text{Noise increase because of sky temperature increases} &= 3 \text{ dB} \\ \text{Total} &= 5.69 \text{ dB} \end{aligned}$$

Thus, the clear sky C/N of 13 dB has been reduced to

$$C/N = 7.31 \text{ dB}$$

which is still above threshold, but getting close. The total carrier-to-noise plus interference ratio is given by

$$\frac{C}{\eta} = \left[\left(\frac{C}{N}\right)^{-1} + \left(\frac{C}{I}\right)^{-1}\right]^{-1}$$

$$\begin{aligned} \frac{C}{\eta} &= \left[\frac{1}{5.38} + \frac{1}{100}\right]^{-1} \\ &= [.186 + .01]^{-1} \\ &= [.196]^{-1} = 5.105 \\ &= 7.1 \text{ dB} \end{aligned}$$

Therefore, C/η is still above threshold. Thus, the system should work more than 99.9 percent of the time.

APPENDIX B

QPSK Modulation and Demodulation

Modulation is a process whereby an information signal modifies a second signal called the carrier. In demodulation, the information signal is recovered from the modulated carrier.

Since most carriers are sinusoids, we can write the carrier as

$$A * \cos[\omega_c * t + \phi]$$

Various modulation schemes have been designed to vary A (amplitude modulation), ω_c (frequency modulation), or ϕ (phase modulation). Varying ω_c or ϕ is sometimes called *angle modulation*. Note that with angle modulation, the amplitude of the modulated carrier does not vary. This leads to the language that an angle-modulated carrier has constant envelope.

In analog modulation, the signals for A, ω_c, or ϕ are usually the information signal or some modification of it. During demodulation, the information signal must be recovered from the modulated carrier, usually in the presence of noise. This recovery of the information signal is mathematically an estimation theory problem.

B.1 Digital Modulation

In digital modulation, the transmitter selects and transmits one of a finite set of messages ($m_1, m_2, \ldots, m_N$). Each message is represented by a unique waveform. Once again, A, ω_c, or ϕ (or combinations thereof) can be varied.

The problem of the demodulator, then, is to select which of the possible messages was most likely to have been sent. Note that this is a decision theory problem.

B.2 The Satellite Channel

The satellite channel, especially above 10 GHz, is characterized by wide variations in amplitude (A). The best solution is to amplify the signal, limit it, and then bandpass the result to recover a fixed-amplitude replica of the original. The limiting process intrinsically eliminates any information carried in A. Thus, high spectral-efficiency techniques such as quadrature amplitude modulation (QAM), which is a combination of amplitude and phase modulation, cannot be used.

In principle, one might design an automatic gain control (AGC) system that would permit an amplitude component. However, this AGC would have to respond in a fraction of a symbol period and operate above 10 GHz. The author is unaware of any system of this type that has been successfully implemented.

B.3 M-ary Phase Shift Keying

M-ary Phase Shift Keying (MPSK) can be represented by

$$A * \cos\left[w_c * t + \frac{k}{M} * 2 * \pi\right],\ k = 0, 1, 2, \ldots, (M - 1)$$

M is normally a power of 2, but doesn't have to be. $M = 2$ is called Binary Phase Shift Keying (BPSK), and $M = 4$ is called Quatenary Phase Shift Keying (QPSK). MPSK for $M = 2$, 4, 8, and 16 is shown in Figures B.1, B.2, B.3, and B.4, the constellations for M-ary PSK.

Note that by using a Grey code for the states, the probability of a bit error is approximately

$$P_b = \frac{P_S}{\log_2[M]}$$

where P_s is the probability of symbol error. This is true because the most likely symbol error is to an adjacent point in the constellation.

B.4 DBS Modulation and Demodulation

All the known DBS systems use QPSK (Figure B.2), so a brief description of QPSK follows.

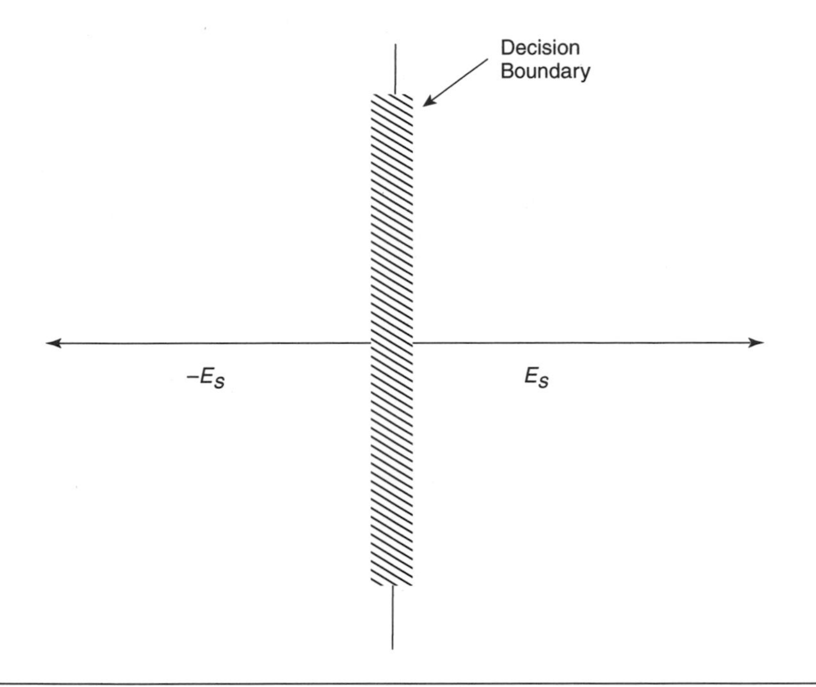

Figure B.1 BPSK

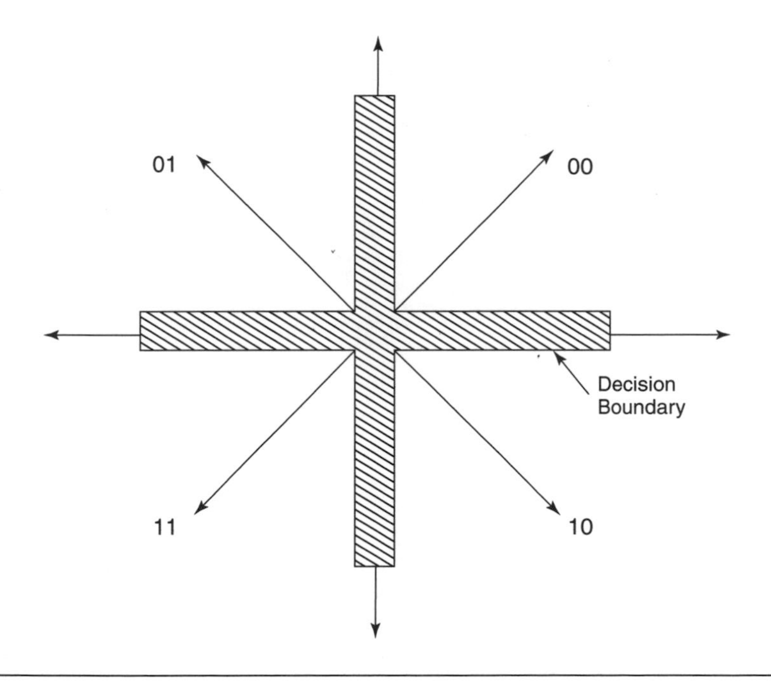

Figure B.2 QPSK

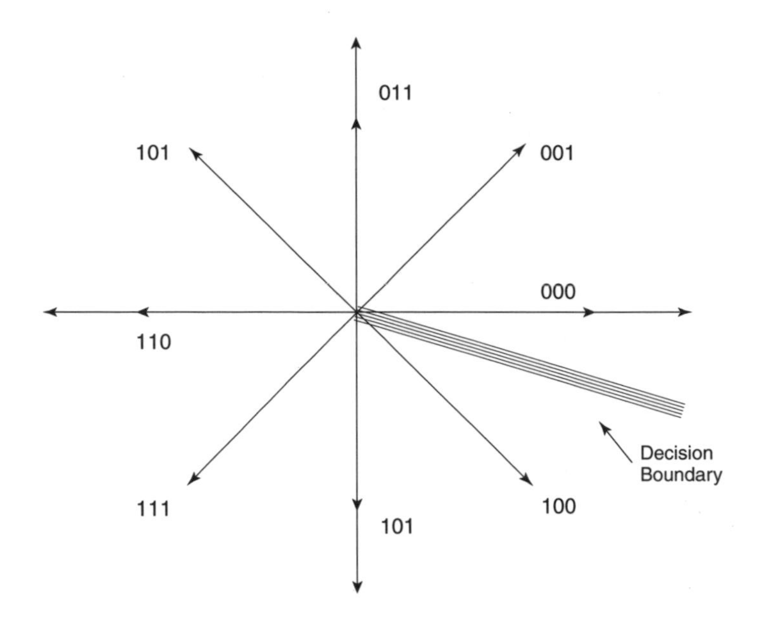

Figure B.3 8 PSK

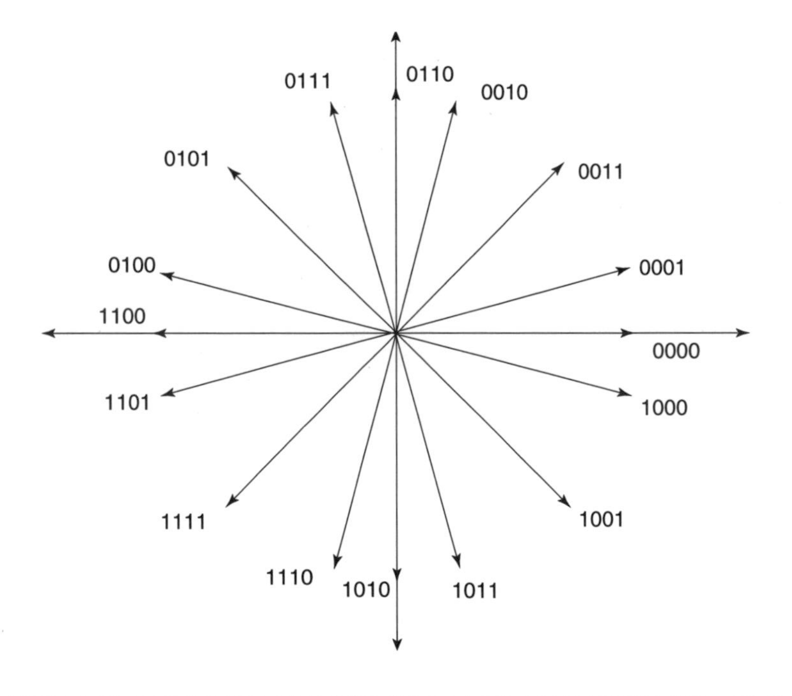

Figure B.4 16 M-ary PSK

B.4.1 Filtering

The Power Spectral Density (PSD) for QPSK is of the form

$$\left[\frac{\sin[\omega_c * t]}{\omega_c * t}\right]^2$$

with the first null at the symbol rate and significant energy extending for multiples of the symbol rate on both sides of the center frequency. This is unacceptable because energy from one transponder will interfere with the transponder on either side of it.

The solution selected by all DBS systems is to filter the baseband signal. The filter of choice is a raised cosine filter of the form

$$P(f) = \begin{cases} 1, & |f| < (1-r) * f_0 \\ \frac{1}{2} * \left(1 + \cos\left[\frac{\pi * \left(|f| - (1-r) * f_0\right)}{2 * r * f_0}\right]\right), & (1-r) * f_0 < |f| < (1+r) * f_0 \\ 0, & |f| > (1+r) * f_0 \end{cases}$$

The parameter r is called the roll-off factor and is 0.35 for all known DBS systems. The frequency f_0 equals the low-pass bandwidth and is usually chosen as 13 MHz for a 24-MHz bandwidth transponder.

B.4.2 Modulation

QPSK can be visualized and implemented by adding two BPSK signals, one of which is rotated 90 degrees with respect to the other. This is shown in Figure B.5.

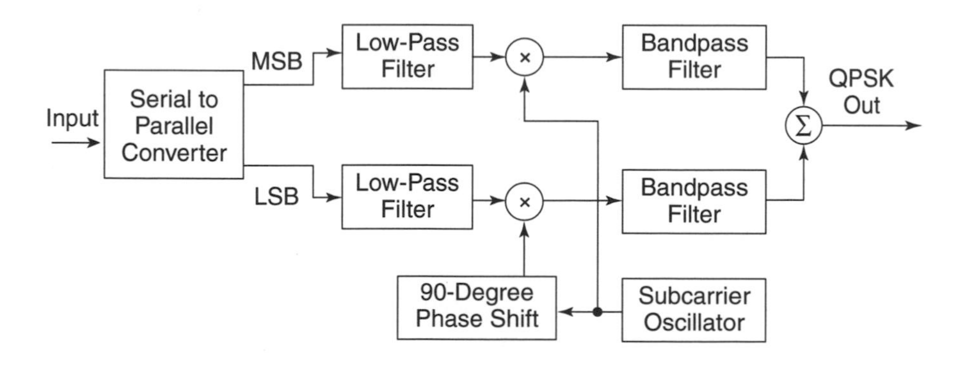

Figure B.5 QPSK Modulator

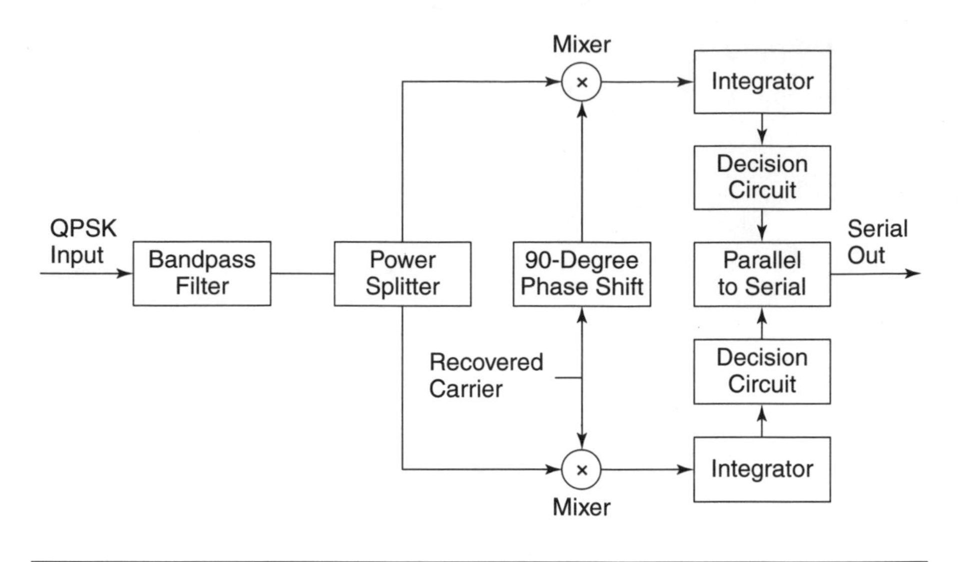

Figure B.6 QPSK Demodulator

B.4.3 Demodulation

The maximum likelihood demodulator is shown in Figure B.6. It should be noted that for soft decision output, analog-to-digital converters must replace the circuits labeled "decision circuit."

APPENDIX C

Algebraic Preliminaries

C.1 Construction of Galois Fields GF(2^m) From GF(2)

Consider adding powers of $\alpha, \alpha^2, \alpha^3, \ldots$ to the binary Galois field with multiplication ($*$) defined as:

$$
\begin{aligned}
0 * 0 &= 0 \\
0 * 1 &= 1 * 0 = 0 \\
1 * 1 &= 1 \\
0 * \alpha &= \alpha * 0 = 0 \\
1 * \alpha &= \alpha * 1 = \alpha \\
\alpha^2 &= \alpha * \alpha \\
\alpha^3 &= \alpha * \alpha * \alpha \\
&\vdots \\
\alpha^j &= \alpha * \alpha * \alpha \ldots \alpha \ (j\, times)
\end{aligned}
$$

It can be shown from this definition that

$$
\begin{aligned}
0 * \alpha^j &= \alpha^j * 0 = 0 \\
1 * \alpha^j &= \alpha^j * 1 = \alpha^j \\
\alpha^j * \alpha^j &= \alpha^j * \alpha^i = \alpha^{i+j}
\end{aligned}
$$

Definition: Degree of a polynomial $p(x)$ is the power of the highest power of x with a nonzero coefficient.

Definition: Irreducible Polynomial—A polynomial $p(x)$ over GF(2) of degree m is irreducible over GF(2) if $p(x)$ is not divisible by any polynomial over GF(2) of degree less than m but greater than zero.

Fact: It can be proved [Lin+83[1]] that any irreducible polynomial over GF(2) of degree m divides $x^{2^m-1} + 1$.

Definition: Primitive Polynomial—An irreducible polynomial $p(x)$ of degree m is primitive if the smallest positive integer η for which $p(x)$ divides $x^{\eta} + 1$ is $\eta = 2^m - 1$.

We can now define a new field F with elements

$$F = \{0, 1, \alpha, \alpha^2, \ldots, \alpha^j, \ldots\}$$

We can then create restrictions on α so that F contains only 2^m elements. Let $p(x)$ be a primitive polynomial of degree m over GF(2). Assume $p(x) = 0$. Since $p(x)$ divides $x^{2^m-1} + 1$, we have

$$x^{2^m-1} + 1 = q(x)p(x)$$

Let $x = \alpha$, so

$$\alpha^{2^m-1} + 1 = q(\alpha)p(\alpha)$$

Because $p(x) = 0$,

$$\alpha^{2^m-1} + 1 = q(x) * 0 = 0$$

$$\therefore \alpha^{2^m-1} = 1$$

Thus, if $p(x) = 0$, the set $F* = \{0, 1, \alpha, \alpha^2, \ldots, \alpha^{2^m-2}]$ is finite and has 2^m elements.

C.2 Basic Properties of the Galois Field GF(2^m)

1. As with real polynomials, a polynomial from GF(2) may not have roots from GF(2). However, whereas polynomials with real coefficients have complex valued roots, a polynomial with coefficients from GF(2) has roots from an extension field of GF(2).

[1] Lin, Shu, and Daniel J. Costello, Jr., *Error Control Coding Fundamentals and Applications*. New York: Prentice-Hall, 1983.

2. If $f(x)$ is a polynomial with coefficients from GF(2) and β is an element of an extension field of GF(2), then β^2 is also a root of $f(x)$.

   Definition: Conjugate of β—The element β^{2^l} is called a conjugate of β.

3. The $2^m - 1$ nonzero elements of $GF(2^m)$ form all the roots of $x^{2^m-1} = 1$. The elements of $GF(2^m)$ form all the roots of $x^{2^m} + x$.

4. Since any element β in $GF(2^m)$ is a root of $x^{2^m} + x$, β may be a root of a polynomial over GF(2) with a degree of less than 2^m. Let $\phi(x)$ be the polynomial of smallest degree over GF(2) such that $\phi(\beta) = 0$.

   Definition: Minimal Polynomial—The polynomial $\phi(x) = y$ is called the minimal polynomial of β.

5. The minimal polynomial $\phi(x)$ of a field element β is irreducible.

6. Let $f(x)$ be a polynomial over GF(2) and $\phi(x)$ be the minimal polynomial of a field element β. If β is a root of $f(x)$, then $f(x)$ is divisible by $\phi(x)$.

7. The minimal polynomial $\phi(x)$ of an element β in $GF(2^m)$ divides $x^{2^m} + x$.

8. Let $f(x)$ be an irreducible polynomial over GF(2) and β be an element in $GF(2^m)$. Let $\phi(x)$ be the minimal polynomial of β. If $f(\beta) = 0$, then $\phi(x) = f(x)$. Thus, β and its conjugates $\beta^2, \beta^{2^2}, \ldots, \beta^{2^{e-1}}$ are roots of $\phi(x)$.

9. Let β be an element in GF(2) and e the smallest integer such that $\beta^{2^e} = \beta$. Then

$$f(x) = \prod_{i=0}^{e-1}(x + \beta^{2^i})$$

   is an irreducible polynomial over GF(2).

10. Let $\phi(x)$ be the minimal polynomial of an element β in $GF(2^m)$. Let e be the smallest integer such that $\beta^{2^e} = \beta$. Then

$$\phi(x) = \prod_{i=0}^{e-1}(x + \beta^{2^i})$$

11. Let $\phi(x)$ be the minimal polynomial of an element β in $GF(2^m)$. Let e be the degree of $\phi(x)$. Then e is the smallest integer such that $\beta^{2^e} = \beta$. Moreover, $e \leq m$.

12. If β is a primitive element of $GF(2^m)$, all of its conjugates are also primitive elements of $GF(2^m)$.

13. If β is an element of order n in $GF(2^m)$, all of its conjugates have the same order n.

APPENDIX D

BCH Code Details

D.1 Generator Polynomial

The generator polynomial of this code is specified in terms of its roots from the Galois field $GF(2^m)$ [Lin+83].[1] Let α be a primitive element in $GF(2^m)$. The generator polynomial $g(x)$ of the t-error correcting BCH code of length $2^m - 1$ is the lowest-degree polynomial over $GF(2)$, which has as its roots:

$$\alpha,\ \alpha^2,\ \alpha^3,\ \ldots,\ \alpha^{2t} \tag{D.1}$$

Thus, $g(x)$ has $\alpha, \alpha^2, \alpha^3, \ldots, \alpha^{2t}$ and their conjugates as all its roots. If $\phi_i(x)$ is the minimal polynomial of α_i, then $g(x)$ is the least common multiple of $\alpha(x)$, $\phi_1(x), \phi_2(x), \ldots, \phi_{2t}(x)$, that is,

$$g(x) = LCM\{\phi_1(x),\ \phi_2(x),\ \ldots,\ \phi_{2t}(x)\} \tag{D.2}$$

It can be shown that the even power of α_i has the same minimum polynomial as some preceding odd power of α in the sequence, and is redundant. Thus, Equation (D.2) can be reduced to

$$g(x) = LCM\{\phi_1(x),\ \phi_2(x),\ \ldots,\ \phi_{2t-1}(x)\} \tag{D.3}$$

Since there are t of these minimal polynomials, each of degree $\leq m$, the degree of $g(x)$ is less than or equal to mt.

Definition: a binary n-tuple

$$v = (v_0,\ v_1,\ \ldots,\ v_{n-1})$$

[1] Lin, Shu, and Daniel J. Costello, Jr., *Error Control Coding Fundamentals and Applications.* New York: Prentice-Hall, 1983.

is a code word if and only if

$$v(x) = v_0 + v_1x + \ldots + v_{n-1}x^{n-1} \tag{D.4}$$

If $v(x)$ has roots from $GF(2)$, $\alpha, \alpha^2, \ldots, \alpha^{2t}$, it follows that $v(x)$ is divisible by the minimal polynomials $\phi_1(x), \phi_2(x), \ldots, \phi_{2t}(x)$ of $d, d^2, \ldots, d^{2t}$ and, hence, by the LCM of the $\phi_i(x)$, $g(x)$. Let

$$v(x) = v_0 + v_1x + \ldots + v_{n-1}x^{n-1}$$

be a code polynomial in a t-error-correcting BCH code of length $n = 2^m - 1$. Since α^i is a root of $v(x)$, $1 \leq i \leq 2t$, then

$$v_0\alpha^i + v_1(\alpha^i) + v_2(\alpha^i)^2 + \ldots v_{n-1}(\alpha^i)^{n-1} = 0$$

This can be written in matrix form as

$$[v_0, v_1, \ldots, v_{n-1}] \begin{bmatrix} 1 \\ \alpha^{2i} \\ \vdots \\ \alpha^{(n-1)i} \end{bmatrix} = 0 \tag{D.5}$$

Next, we can form the matrix

$$H = \begin{bmatrix} \alpha & \alpha^2 & \ldots & \alpha^{\eta-1} \\ \alpha^2 & (\alpha^2)^2 & \ldots & (\alpha^2)^{\eta-1} \\ \alpha^3 & (\alpha^3)^2 & \ldots & (\alpha^3)^{\eta-1} \\ & & \vdots & \\ (\alpha^{2*T}) & (\alpha^{2*T})^2 & \ldots & (\alpha^{2*T})^{\eta-1} \end{bmatrix} \tag{D.6}$$

Thus, the code is the null space of the matrix H, and H is the parity-check matrix of the code. The matrix H can be reduced so that the basic entries are odd powers of α.

It can be shown [Lin+83] that the minimum weight of the t-error-correcting code just defined is at least $2t + 1$. Thus, the minimum distance of the code is at least $2t + 1$, which is called the designed distance of the code. The true distance of a particular code may be equal to or greater than the designed distance.

D.1.1 Decoding of BCH Codes

Suppose that a code word

$$v(x) = v_0 + v_1x + \ldots v_{n-1}x^{n-1}$$

is transmitted and the vector

$$r(x) = r_0 + r_1 x + \dots + r_{n-1} x^{n-1}$$

is received. If $e(x)$ is the error pattern, then

$$r(x) = v(x) + e(x)$$

Three steps are required to correct $r(x)$:

1. Complete the syndrome, $s = (s_1, s_2, \dots, s_{2t})$
2. Determine the error location polynomial $\sigma(x)$ from the syndrome components $s_1, s_2, \dots, s_{2t}$.
3. Determine the error location by numbers $\beta_1, \beta_2, \dots, \beta_v$ by finding the roots of $\sigma(x)$ and correct the errors in $r(x)$.

These are discussed in detail in the following sections.

Step 1 Compute the Syndrome

$$\begin{aligned} \mathbf{S} &= (s_1, s_2, \dots, s_{\partial t}) = \mathbf{r} \cdot \mathbf{H}^T \\ S_i &= \mathbf{r}(\alpha^i) = r_0 + r_1\alpha^i + \dots r_{n-1}(\alpha^i)^{(n-1)} \end{aligned} \quad \text{(D.7)}$$

To calculate the s_i, divide $r(x)$ by the minimal polynomial $\phi_i(x)$ of α^i to obtain

$$\mathbf{r}(x) = a_i(x)\,\phi_i(x) + \mathbf{b}_i(x)$$

where $b_i(x)$ is the remainder with degree less than $\phi_i(x)$. Since $\phi_i(\alpha^i) = 0$, we have

$$S_i = \mathbf{r}(\alpha^i) = \mathbf{b}_i(\alpha^i)$$

Since $\alpha^1, \alpha^2, \dots, \alpha^{2t}$ are roots of the code polynomial, $\mathbf{V}(\alpha^i) = 0$ for $1 \le i \le 2t$, so

$$S_i = \mathbf{e}(\alpha^i)$$

If the error pattern $\mathbf{e}(x)$ has ν error locations $x^{j1}, x^{j2}, ..., x^{j\nu}$, so that

$$\mathbf{e}(x) = x^{j1} + x^{j2} + \dots + x^{j\nu} \quad \text{(D.8)}$$

we obtain the following set of equations:

$$\begin{aligned} s_1 &= \alpha^{j1} + \alpha^{j2} + \cdots + \alpha^{j\nu} \\ s_2 &= (\alpha^{j1})^2 + (\alpha^{j2})^2 + \dots + (\alpha^{j\nu})^2 \\ &\vdots \\ s_{2t} &= (\alpha^{j1})^{2t} + (\alpha^{j2})^{2t} + \dots + (\alpha^{j\nu})^{2t} \end{aligned} \quad \text{(D.9)}$$

Any method of decoding the set of Equation (D.9) is a decoding algorithm for the BCH codes. There are $2k$ solutions for Equation (D.9). If the number of errors in $\mathbf{e}(x)$ is less than or equal to t, the solution with the smallest number of errors is correct.

Step 2 Iterative Algorithm for Finding the Error-Location Polynominal $\sigma(x)$

For convenience in notation, let $\beta_1 = \alpha^{jl}$ so that the set of Equations (D.9) becomes

$$\begin{aligned} s_1 &= \beta_1 + \beta_2 + \dots + \beta_\nu \\ s_2 &= \beta_1^2 + \beta_2^2 + \dots + \beta_\nu^2 \\ &\vdots \\ s_{2t} &= \beta_1^{2t} + \beta_2^{2t} + \dots + \beta_\nu^{2t} \end{aligned} \quad \text{(D.10)}$$

These $2t$ equations are symmetric functions in $\beta_1, \beta_2, ..., \beta_\nu$ that are known as power–sum symmetric functions.

Next, define $\sigma(x)$ as

$$\begin{aligned} \sigma(x) &= (1 + \beta_1 x)(1 + \beta_2 x) \dots (1 + \beta_\nu x) \\ &= \sigma_0 + \sigma_1 x + \sigma_2 x^2 + \dots + \sigma_\nu x^\nu \end{aligned} \quad \text{(D.11)}$$

The roots of $\sigma(x)$, $\beta_1, \beta_2, ..., \beta_\nu$ are the inverses of the error-location numbers. For this reason, $\sigma(x)$ is called the error-location polynomial.

The coefficients of $\sigma(x)$ and the error-location numbers are related by

$$\begin{aligned} \sigma_0 &= 1 \\ \sigma_1 &= \beta_1 + \beta_2 + \dots \beta_\nu \\ \sigma_2 &= \beta_1\beta_2 + \beta_2\beta_3 + \dots \beta_{\nu-1}\beta_\nu \\ &\vdots \\ \sigma_\nu &= \beta_1\beta_2 + \dots \beta_\nu \end{aligned} \quad \text{(D.12)}$$

The σ_i are called elementary symmetric functions of the β_1s. The σ_i's are related to the s_j by

$$
\begin{aligned}
s_1 + \sigma_1 &= 0 \\
s_2 + \sigma_1 s_1 + 2\sigma_2 &= 0 \\
s_3 + \sigma_1 s_2 + \sigma_2 s_1 + 3\sigma_3 &= 0 \\
&\vdots \\
s_v + \sigma_1 s_{v-1} + \dots \sigma_{v-1} s_1 + v\sigma_v &= 0 \\
s_{v+1} + \sigma_1 s_v + \dots \sigma_{v-1} s_2 + \sigma_v s_1 &= 0
\end{aligned}
\tag{D.13}
$$

These are called Newton's identities. These must be solved for the $\sigma(x)$ of minimum degree.

Step 3 Finding the Error-Location Numbers and Error Correction

The last step in decoding a BCH code is to find the error-location numbers that are the reciprocals of the roots of $\sigma(x)$. The roots of $\sigma(x)$ can be found by substituting $1, \alpha, \alpha^2, \dots, \alpha^{\eta-1}$ into $\sigma(x)$. If α^l is a root of $\sigma(x)$, α^{n-l} is an error-location number and the received digit r_{n-l} is an erroneous digit.

D.1.2 Iterative Procedure (Berlekamp's Algorithm)

The iteration number is a superscript, $\sigma^{(\mu)}(x)$. The first iteration is to find a minimum-degree polynomial $\sigma^{(1)}(x)$ whose coefficients satisfy the first Newton identity.

For the second iteration, check if $\sigma^{(1)}(x)$ satisfies the second Newton identity. If it does, $\sigma^{(2)}(x) = \sigma^{(1)}(x)$. If the coefficients of $\sigma^{(1)}(x)$ do not satisfy the second Newton identity, a correction term is added to $\sigma^{(1)}(x)$ to form $\sigma^{(2)}(x)$ such that $\sigma^{(2)}(x)$ has minimum degree and its coefficients satisfy the first two Newton identities.

This procedure is continued until $\sigma^{(2t)}(x)$ is obtained. Then,

$$\sigma(x) = \sigma^{(2t)}(x)$$

To calculate the discrepancy, let

$$\sigma^{(\mu)}(x) = 1 + \sigma_1 x^{(\mu)} + \sigma_2^{(\mu)} x^2 + \dots + \sigma_{l\mu}^{(\mu)} x^{l\mu}$$

Table D.1 Iterative Algorithm for $\sigma^{(\mu)}(x)$

μ	$\sigma^{(\mu)}(x)$	d_μ	ι_μ	$\mu - \iota_\mu$
–1	1	1	0	–1
0	1	s_1	0	0
1				
2				
⋮				
$2t$				

be the minimum-degree polynomial determined at the μth iteration whose coefficients satisfy the first μ Newton identities. To determine $\sigma^{(\mu+1)}(x)$ we compute the μth discrepancy:

$$d_\mu = s_{\mu+1} + \sigma_1^{(\mu)} s_\mu + \sigma_2^{(\mu)} s_{\mu-1} + \ldots + \sigma_{l_\mu}^{(\mu)} s_{\mu+1-l_\mu}$$

If $d_\mu = 0$, then $\sigma^{(\mu+1)}(x) = \sigma^{(\mu)}(x)$. However, if $d_\mu \neq 0$, go back to the iteration number less than μ and determine a polynomial $\sigma^{\rho}(x)$ such that the ρth discrepancy $d_\rho \neq 0$ and $\rho - \iota_\rho$ has the largest value. Then,

$$\sigma^{(\mu+1)}(x) = \sigma^{(\mu)}(x) + d_\mu d_\rho^{-1} x^{(\mu-\rho)} \sigma^{(\rho)}(x) \tag{D.14}$$

To carry out the iterations of finding $\sigma(x)$, we begin with Table D.1 and fill out the rows where ι_μ is the degree of $\sigma^{(\mu)}(x)$.

If we have filled Table D.1 through the μth row, the $\mu + 1$ row is calculated as follows:

1. If $d_\mu = 0$, then $\sigma^{(\mu+1)}(x) = \sigma^{(\mu)}(x)$ and $\iota_{\mu+1} = \iota_\mu$.
2. If $d_\mu \neq 0$, find another row ρ prior to the μth row such that $d_\rho \neq 0$ and the number $\rho - \iota_\rho$ in the last column of the table has the largest value. Then $\sigma^{(\mu+1)}(x)$ is given by Equation (D.14) and

$$\iota_{\mu+1} = \max(\iota_\mu, \iota_\mu + \mu - \rho)$$

3. In either case,

$$d_{\mu+1} = s_{\mu+2} + \sigma_{\iota}^{(\mu+1)} s_{\mu+1} + \ldots + \sigma_{\iota_{\mu+1}}^{\mu+1} s_{\mu+2-\iota_{\mu+1}}$$

If the number of errors in the received polynomial is less than the designed error-correcting capability t of the code, it is not necessary to carry out the $2t$ steps for finding $\sigma(x)$. It can be shown that if $\sigma^{(\mu)}(x)$ and if d_{μ} and the discrepancies at the next $t - \iota_{\mu} - 1$ iterations are 0, the solution is $\sigma^{(\mu)}(x)$.

While this is interesting and can reduce the average computation time, it leads to variable rate decoding. Since decoding a particular vector may take all $2t$ steps, most systems will be designed to allow this amount of time for decoding.

APPENDIX E

Cyclical Redundancy Code

In most packetized communication systems, it can be valuable to know if a transmission error has occurred. If a return link is available, a Negative Acknowledge (NAK) can be sent back to the transmitter so the packet can be resent. In broadcast audio/visual services, it is usually possible to conceal errors if it is known that an error has occurred. Thus, both the Systems and Audio parts of MPEG incorporate a Cyclical Redundancy Code (CRC) to detect errors in the bitstream.

In Chapter 5, simple Elias codes were introduced. It was shown that by breaking a message into 4-bit nibbles and calculating row- and column-parity checks that were appended to the message, all single bit errors in the message could be corrected. If only the column-parity bits were calculated and appended to the message, all single bit errors in the message could be detected. The CRC provides this type of error detection and operates in the same way: A set of parity bits is calculated for the message and appended to the message bits. However, as we shall see, the CRC is much more powerful than simple parity checks.

The problem with simple parity checks is that they are based on addition. Each incoming byte affects only one byte in the parity bits. It turns out that a form of division can be used to calculate the CRC, which is much more powerful. It can be shown that the probability of an undetected error for the CRC-32 (32 bits) is $1/2^{32}$.

The idea behind the CRC is to treat the message as a very large binary number and divide it by another binary number. The quotient is thrown away and the remainder is used as the set of parity bits, sometimes called the *checksum*. The divisor is usually presented in the form of a polynomial called the generating polynomial. For the 16-bit CRC used for MPEG 1 Audio, the generator polynomial is $G(x) = x^{16} + x^{15} + x^2 + 1$. Basically, this means

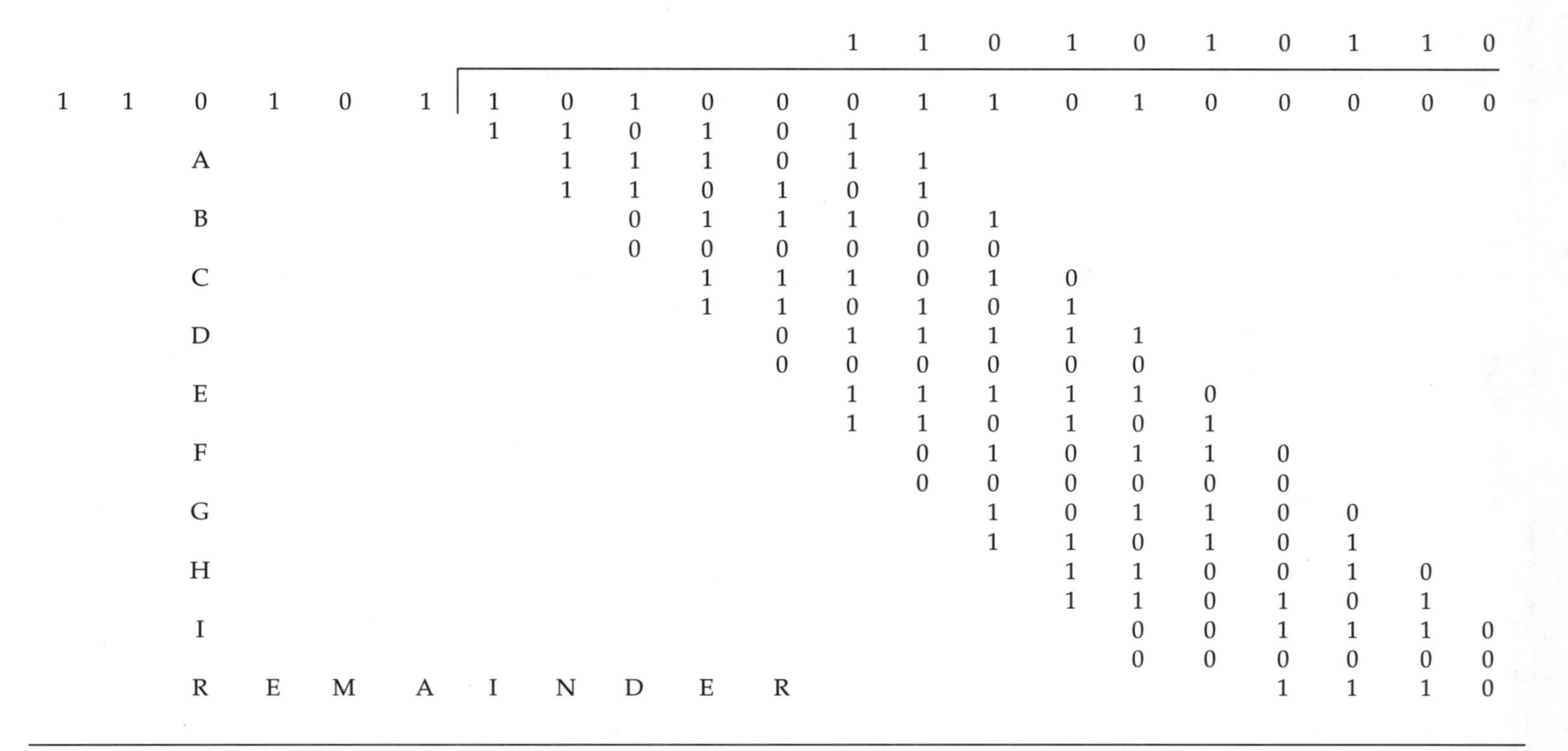

Figure E.1 Division to Form CRC

that the divisor has a '1', where the power of x is nonzero and zeroes elsewhere.

The division to form the CRC parity bits is a bit unusual. First, it uses the following binary arithmetic:

$$0 \pm 0 = 0$$
$$1 \pm 0 = 1$$
$$0 \pm 1 = 1$$
$$1 \pm 1 = 0$$

Also, when performing the division, if the part of the dividend that is being divided by the divisor for a particular quotient bit starts with a '1', the quotient bit is a '1'; if '0', the quotient bit is a '0'.

This is illustrated by the following example.

Generator polynomial:

$$x^5 + x^4 + x^2 + 1$$

which can be represented by '110101'.

Message: '1010001101'. The generating polynomial uses 6 binary positions. One less than this, or 5 zeroes, are appended to the message to form the dividend. The division is then shown in Figure E.1.

Several items about the figure should be noted. First, the column of capital letters—A through I on the left—will be used later. Next, note the first step. The quotient bit is a '1', even though '110101' is larger than '101000' because '101000' starts with a '1'. Finally, the line labeled REMAINDER is the CRC value, '1110'. The message transmitted then is '10100011011110'—the original message with the 4 CRC bits appended.

In the decoder, the message bits are divided by the same divisor. If the result is '1110', the probability is very high that there have been no errors.

One of the interesting aspects of the CRC is that the division to find the CRC value can be performed by a shift register and Exclusive OR circuits. Consider the circuit of Figure E.2. Note in the figure that the Exclusive OR circuits are associated with the outputs of the fifth stage (the one at the input), the fourth stage, and the second. These are the powers of x in the generating polynomial that are nonzero. The stages are flip-flops or delay elements. Table E.1 shows the process as the input message (shown in the right

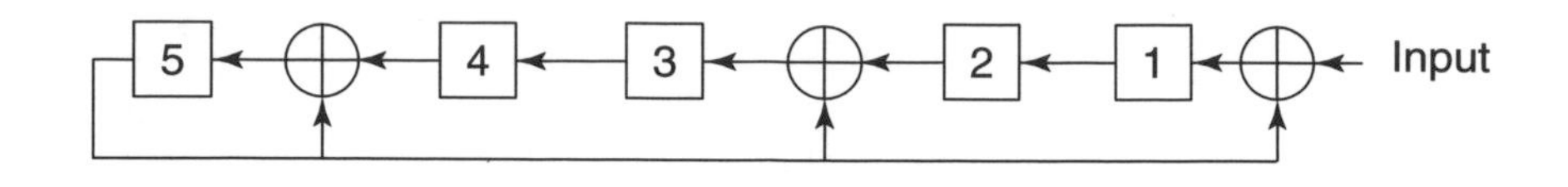

Figure E.2 Shift Register Implementation of CRC Example

column) is shifted into the shift register. The register starts with all zeroes. The output of each stage after the shift is the value of its input before the shift. Compare the rows labeled A through I in the table with the steps in the division in Figure E.1. They are the same. Thus, the simple shift register is implementing the division.

An obvious question is "What makes a good generator polynomial?" Perhaps the best answer is that the CRCs have been around a long time and have been studied extensively. Some well-known standards have chosen specific generating polynomials.

Table E.1 Steps in Shift Register Generation of CRC

Shift	Reference	Stage 5	Stage 4	Stage 3	Stage 2	Stage 1	Input
Start		0	0	0	0	0	1
1		0	0	0	0	1	0
2		0	0	0	1	0	1
3		0	0	1	0	1	0
4		0	1	0	1	0	0
5		1	0	1	0	0	0
6	A	1	1	1	0	1	1
7	B	0	1	1	1	0	1
8	C	1	1	1	0	1	0
9	D	0	1	1	1	1	1
10	E	1	1	1	1	1	0
11	F	0	1	0	1	1	0
12	G	1	0	1	1	0	0
13	H	1	1	0	0	1	0
14	I	0	0	1	1	1	0
15			1	1	1	0	

$$CRC\text{-}12 = x^{12} + x^{11} + x^{3} + x^{2} + 1$$

$$CRC\text{-}16 = x^{16} + x^{15} + x^{2} + 1$$

$$CRC\text{-}CCITT = x^{16} + x^{12} + x^{5} + 1$$

$$\text{Ethernet} = x^{31} + x^{26} + x^{22} + x^{16} + x^{12} + x^{11} + x^{10} + x^{8} + x^{7} + x^{5} + x^{4}s + x^{2} + x + 1$$

MPEG 2 Systems utilizes the 32-bit Ethernet CRC. In MPEG 1 Audio, the 16-bit CRC-16 is used. The protected fields are bits 16 to 31 of the header, the bit allocation, and scalefactor selection information.

APPENDIX F

A5 Matrix

Table F.1 A5 Matrix

$$
\begin{bmatrix} v_{0001} \\ v_{0010} \\ v_{0011} \\ v_{0100} \\ v_{0101} \\ v_{0110} \\ v_{0111} \\ v_{1000} \\ v_{1001} \\ v_{1010} \\ v_{1011} \\ v_{1100} \\ v_{1101} \\ v_{1110} \\ v_{1111} \end{bmatrix}
\begin{bmatrix}
0 & 0 & 0 & 0 & 0 & 0 & 0 & I & 0 & 0 & 0 & 0 & 0 & 0 & 0 \\
W & 0 & 0 & 0 & 0 & 0 & 0 & 0 & W & 0 & 0 & 0 & 0 & 0 & 0 \\
I*W & 0 & 0 & 0 & 0 & 0 & 0 & 0 & I*W & 0 & 0 & 0 & 0 & 0 & 0 \\
0 & \mathrm{WI} & 0 & 0 & 0 & 0 & 0 & 0 & W & 0 & 0 & 0 & 0 & 0 & 0 \\
0 & I*W & 0 & 0 & 0 & 0 & 0 & 0 & 0 & I*W & 0 & 0 & 0 & 0 & 0 \\
0 & 0 & I & 0 & 0 & 0 & 0 & 0 & 0 & 0 & W^2 & 0 & 0 & 0 & 0 \\
0 & 0 & I*W^2 & 0 & 0 & 0 & 0 & 0 & 0 & 0 & I & 0 & 0 & 0 & 0 \\
0 & 0 & 0 & W & 0 & 0 & 0 & 0 & 0 & 0 & 0 & W & 0 & 0 & 0 \\
0 & 0 & 0 & I*W & 0 & 0 & 0 & 0 & 0 & 0 & 0 & I*W & 0 & 0 & 0 \\
0 & 0 & 0 & 0 & W^2 & 0 & 0 & 0 & 0 & 0 & 0 & 0 & I & 0 & 0 \\
0 & 0 & 0 & 0 & I & 0 & 0 & 0 & 0 & 0 & 0 & 0 & I*W & 0 & 0 \\
0 & 0 & 0 & 0 & 0 & W^2 & 0 & I & 0 & 0 & 0 & 0 & 0 & 1 & 0 \\
0 & 0 & 0 & 0 & 0 & I & 0 & 0 & 0 & 0 & 0 & 0 & 0 & I*W^2 & 0 \\
0 & 0 & 0 & 0 & 0 & 0 & W & 0 & 0 & 0 & 0 & 0 & 0 & 0 & W \\
0 & 0 & 0 & 0 & 0 & 0 & I*W & 0 & 0 & 0 & 0 & 0 & 0 & 0 & I*W
\end{bmatrix}
\begin{bmatrix} v_{0001} \\ v_{0010} \\ v_{0011} \\ v_{0100} \\ v_{0101} \\ v_{0110} \\ v_{0111} \\ v_{1000} \\ v_{1001} \\ v_{1010} \\ v_{1011} \\ v_{1100} \\ v_{1101} \\ v_{1110} \\ v_{1111} \end{bmatrix}
$$

APPENDIX G

Operators, Mnemonics, Abbreviations

The mathematical operators used in this book are similar to those used in the C programming language. However, integer division with truncation and rounding are specifically defined. The bitwise operators are defined assuming two's-complement representation of integers. Numbering and counting loops generally begin from 0.

Operators

Mathematical

Operator	Meaning
+	Addition
–	Subtraction (as a binary operator) or negation (as a unary operator)
++	Increment
--	Decrement
*	Multiplication
^	Power
/	Integer division with truncation of the result toward 0. For example, 7/4 and –7/–4 are truncated to 1 and –7/4 and 7/–4 are truncated to –1.
//	Integer division with rounding to the nearest integer. Half-integer values are rounded away from 0 unless otherwise specified. For example, 3//2 is rounded to 2 and –3//2 is rounded to –2.
DIV	Integer division with truncation of the result toward $-\infty$.
%	Modulus operator; defined only for positive numbers.

Sign( )	Sign(x) = 1 x > 0 0 x == 0 –1 x < 0
NINT( )	Nearest integer operator. Returns the nearest integer value to the real-valued argument. Half-integer values are rounded away from 0.
sin	Sine
cos	Cosine
exp	Exponential
√	Square root
$\log_{10}$	Logarithm to base 10
$\log_e$	Logarithm to base *e*

Logical

||	Logical OR
&&	Logical AND
!	Logical NOT

Relational

>	Greater than
≥	Greater than or equal to
<	Less than
≤	Less than or equal to
==	Equal to
!=	Not equal to
max [,...,]	The maximum value in the argument list

Bitwise and Assignment

&	AND
|	OR
>>	Shift right with sign extension
<<	Shift left with 0 fill
=	Assignment operator

Mnemonics

The following mnemonics are defined to describe the different data types used in the coded bitstream.

bslbf	Bit string, left bit first, where "left" is the order in which bit strings are written in the Recommendation I International Standard. Bit strings are written as a string of 1s and 0s within single quote marks (e.g., '1000 0001'). Blanks within a bit string are for ease of reading and have no significance.
ch	Channel
gr	Granule of 3 * 32 subband samples in Audio Layer II, 18 * 32 subband samples in Audio Layer III.
main_data	The **main_data** portion of the bitstream contains the scalefactors, Huffman encoded data, and ancillary information.
main_data_beg	This gives the location in the bitstream of the beginning of the **main_data** for the frame. The location is equal to the ending location of the previous frame's **main_data** plus 1 bit. It is calculated from the **main_data_end** value of the previous frame.
part 2_length	This value contains the number of **main_data** bits used for scalefactors.
rpchof	Remainder polynomial coefficients, highest order first.
sb	Subband
scfsi	Scalefactor selector information
switch_point_1	The number of the scalefactor band (long block scalefactor band) from which point on window switching is used.
switch_point_s	The number of the scalefactor band (short block scalefactor band) from which point on window switching is used.
tcimsbf	Two's complement integer, msb (sign) bit first.
uimsbf	Unsigned integer, most significant bit first.
vlclbf	Variable length code, left bit first, where "left" refers to the order in which the VLC codes are written.
window	The number of an actual time slot in the case of **block_type==2**, $0 \leq$ window ≤ 2.

The byte order of multibyte words is most significant byte first.

Constants

π	3.14159265359
e	2.71828182845

Method of Describing Bitstream Syntax

The action caused by a decoded data element in a bitstream depends on the value of that data element and on data elements previously decoded. The decoding of the data elements and definition of the state variables used in their decoding are described in the clauses containing the semantic description of the syntax. The following constructs are used to express the conditions when data elements are present and are in normal type.[1]

Construct	Description
while (condition) {	If the condition is true, the group of data elements occurs next in the data stream. This repeats until the condition is not true.
data_element	The data element always occurs at least once.
...	
}	
do {	The data element is repeated until the condition is not true.
data_element	
...	
} while (condition)	
if (condition) {	If the condition is true, the first group of data elements occurs next in **data_element** in the data stream.
	...
else {	If the condition is not true, the second group of data elements occurs next in **data_element** in the data stream.
...	
}	
for (i = 0;i < n; i++) {	The group of data elements occurs *n* times. Conditional constructs within the group of data elements may depend on the value of the loop control variable *i*, which is set to 0 for the first occurrence, incremented to 1 for the second occurrence, and so forth.
data_element	
...	
}	

[1] *Note:* This syntax uses the "C"-code convention that a variable or expression evaluating to a nonzero value is equivalent to a condition that is true.

As noted, the group of data elements may contain nested conditional constructs. For compactness, the { } are omitted when only one data element follows.

data_element [ ] An array of data. The number of data elements is indicated by the context.

data_element [n] The $n + 1^{th}$ element of an array of data.

data_element [m][n] The $m + 1, n + 1^{th}$ element of a two-dimensional array of data.

data_element[1][m][n] The $1 + 1, m + 1, n + 1^{th}$ element of a three-dimensional array of data.

data_element[m...n] The inclusive range of bits between bit m and bit n in the **data_element.**

Function Definitions

bytealigned( ) This function returns '1' if the current position is on a byte boundary; that is, the next bit in the bitstream is the first bit in a byte. Otherwise, it returns '0'.

nextbits( ) This function permits comparison of a bit string with the next bits to be decoded in the bitstream.

next_start_code This function removes any 0 bit and 0 byte stuffing and locates the next start code. The function checks whether the current position is byte aligned. If it is not, 0 stuffing bits are present. After that, any number of 0 bytes may be present before the **start_code**. Therefore, start codes are always byte aligned and may be preceded by any number of 0 stuffing bits.

next_start_code Table

Syntax	Number of bits	Mnemonic
next_start_code() {		
while (!bytealigned())		
zero_bit	1	'0'
while (nextbits() ! = '0000 0000 0000 0000 0000 0001')		
zero_byte	8	'00000000'
}		

GLOSSARY

access unit [system]: A coded representation of a presentation unit. In the case of compressed audio, an access unit is an Audio Access Unit. In the case of compressed video, an access unit is the coded representation of a picture.

authentication: Any technique enabling the receiver to automatically identify and reject messages that have been altered deliberately or by channel errors.

bit rate: The rate at which the compressed bitstream is delivered from the channel to the input of a decoder.

block code: A code in which parity bits are added to the message bits.

byte-aligned: A bit in a coded bitstream is byte-aligned if its position is a multiple of 8-bits from the first bit in the stream.

certificates: A nonforgeable, tamper-proof way of certifying the validity and, therefore, protecting the integrity of published public keys. Digital certificates are issued, verified, and revoked by a certifying authority, which can be any trusted party for central administration.

channel: A digital medium that stores or transports a bitstream.

coded representation: A data element as represented in its encoded form.

code rate: In a coded bitstream, the ratio of the information bits to the total bits.

compression: A reduction in the number of bits used to represent an item of data.

conditional access: The control mechanisms, data structures, and commands that provide for selective access and denial of specific services. Conditional access (CA) systems use signal security, which is any technology, such as encryption, that can prevent a signal from being received by anyone except authorized users.

constant bit rate: An operation in which the bit rate is constant from the start to the finish of the compressed bitstream.

control word: The frequency varying key used to decrypt a service. Control words are delivered in ECMs.

cryptography: The art or process of writing or deciphering secret code.

Cyclic Redundancy Code (CRC) check: A way to verify the correctness of data.

data element: An item of data before encoding and after decoding.

data encryption standard (DES): A block ciphering technique described in a U.S. government standard (FIPS PUB 46-2) used for privacy protection. It is probably the best-known and most widely used crypto-algorithm in the world. As a secret-key, symmetric system, it requires the exchange of secret encryption keys between users.

decoded stream: The decoded reconstruction of a compressed bitstream.

decoder: An embodiment of a decoding process.

decoding (process): The process-defined MPEG standard that reads an input-coded bitstream and outputs decoded pictures or audio samples.

decoding time stamp (DTS) [system]: A field that may be present in a PES packet header that indicates the time in which an access unit is decoded in the system target decoder.

demodulator (QPSK): Selects which of the four possible subcarrier phases was transmitted. It produces a 2-bit digital output.

digital signature: An electronic "passport" that only one entity can produce, but all others can verify with the sender's public key. Digital signatures provide an unambiguous confirmation of the identity of the sender of a message. With digital signatures, messages from unauthorized sources can be rejected and authorized messages cannot be repudiated by the sender.

digital storage media (DSM): A digital storage or transmission device or system.

DIRECTV: The subsidiary of Hughes Electronics that provides the service of the same name.

Early Bird: The first commercial communication satellite. Developed by Hughes Aircraft Company for COMSAT. Its official name is INTELSAT I.

editing: The process by which one or more compressed bitstreams are manipulated to produce a new compressed bitstream. Conforming edited bitstreams must meet the requirements defined in the MPEG standard.

elementary stream (ES) [system]: A generic term for one of the coded video, coded audio, or other coded bitstreams.

Elementary Stream Clock Reference (ESCR): A time stamp in the PES Stream from which decoder timing is derived.

encoder: An embodiment of an encoding process.

encoding (process): A process, not specified in the MPEG standard, that reads a stream of input pictures or audio samples and produces a valid coded bitstream as defined in the MPEG standard.

encryption: Transforms digital data into a format that is unintelligible without the proper key for decryption. Sometimes known as "scrambling" when applied to a digital service. The design of the algorithm and the key length employed will determine the strength of the cryptographic protection. Encryption can be used both for service denial/access and message privacy.

energy dispersal: The use of a pseudorandom generator technique to ensure that the bitstream does not contain runs of the same type of bit.

entitlement: Data structures transmitted within messages having cryptographic protections that authorize reception of a service at a set-top terminal. Entitlements may be delivered in EMMs via the MPEG 2 Transport Stream or via "out of band" communication channels.

entitlement control message (ECM): Messages containing systemwide information that "unlocks" an encrypted service by transmitting control words. Each ECM, which is unique for each service, enables cryptographic partitioning so that different service providers can selectively grant access to their own services.

entitlement management message (EMM): Messages that are individually addressed to or sent by specific receivers that allow delivery of secure and authenticated authorizations and supports nonrepudiation of service orders. RSA's public key cryptography typically will be used for EMM transmission.

entropy coding: Variable length lossless coding of the digital representation of a signal to reduce redundancy.

event: A collection of Elementary Streams with a common time base, an associated start time, and an associated end time.

fast forward playback [video]: The process of displaying a sequence or parts of a sequence of pictures in display-order faster than real time

Federal Communications Commission (FCC): The U.S. government agency responsible for regulating communications.

fingerprinting: Methods used in a signal to identify the source decoder, thereby providing a mechanism to detect and trace unauthorized copying of material.

forbidden: When used in the clauses defining the coded bitstream, indicates that the value shall never be used. This is usually to avoid emulation of start codes.

Forward Error Correction: Use of parity bits and codes to reduce transmission errors.

HS 601: Hughes Electronics satellite family that is the basis for the DBS satellites.

Integrated Receiver Decoder (IRD): The indoor part of a DBS receiving unit.

Kepler's Laws: Rules, which describe what governs the motion of bodies orbiting the Earth, developed by Johannes Kepler.

layer [Video and Systems]: One of the levels in the data hierarchy of the Video and Systems specifications defined in Parts 1 and 2 of the MPEG 2 standard.

low-noise block (LNB): Part of a receiver's outdoor unit that amplifies an incoming K_u signal and then downconverts it into L-Band.

pack [system]: Consists of a pack header followed by zero or more packets. It is a layer in the system coding syntax.

packet [system]: Contiguous bytes of data from an Elementary Stream present in a packet.

packet data [system]: Consists of a header followed by a number of contiguous bytes from an elementary data stream. It is a layer in the system coding syntax.

packet identifier (PID) [system]: A unique integer value used to associate Elementary Streams of a program in a single or multiprogram Transport Stream.

Packetized Elementary Stream (PES) [system]: The data structure used to carry Elementary Stream data. It consists of a PES packet header followed by PES packet payload.

padding [audio]: A method to adjust the average length of an audio frame in time to the duration of the corresponding PCM samples by conditionally adding a slot to the audio frame.

payload: The bytes that follow the header bytes in a packet. For example, the payload of a Transport Stream packet includes the **PES_packet_header** and its **PES_packet_data_bytes, pointer_field** and PSI sections, or **private_data**, but a

PES_packet_payload consists of only **PES_packet_data_bytes**. The Transport Stream packet header and adaptation fields are not payload.

PES [system]: An abbreviation for Packetized Elementary Stream.

PES Stream [system]: Consists of PES packets, all of whose payloads consist of data from a single Elementary Stream and all of which have the same **stream_id**. Specific semantic constraints apply.

physical security: Methods used to thwart piracy and prevent unauthorized access to crucial elements of a CA system. In a set-top terminal, a secure microprocessor supports nonvolatile storage of keys and authorizations, and assures a protected environment for decryption and encryption, authentication, and access control logic.

piracy: Any impersonation, unauthorized browsing, falsification or theft of data, or disruption of service or control information in a network.

presentation time stamp (PTS) [system]: A field that may be present in a PES packet header indicating the time that a presentation unit is presented in the system target decoder.

presentation unit (PU) [system]: A decoded Audio Access Unit or a decoded picture.

private key: The decryption (reception) or the encryption (signature) component of an asymmetric key set.

program [system]: A collection of Elementary Streams with a common timebase and intended for synchronized presentation.

Program Clock Reference (PCR) [system]: A time stamp in the Transport Stream from which decoder timing is derived.

Program Specific Information (PSI) [system]: Consists of normative data that is necessary for the demultiplexing of Transport Streams and the successful regeneration of programs. One case of PSI, the nonmandatory Network Information Table, is privately defined.

protocols: Specific rules, procedures, or conventions relating to the format and timing of data transmission between two devices. For broadband networks, international standards such as MPEG, DVB, and DAVIC define appropriate protocols.

public key: Algorithms that encrypt and decrypt by using asymmetric (different), yet mathematically linked keys. Each security module is assigned a pair of keys: The encryption key is "public" and does not require distribution by secure means. The decryption, or "private," key cannot be discovered through knowledge of the public key or its underlying algorithm. Public key algorithms can apply to one or more of the following: key distribution, encryption, authentication, or digital signature.

Quadrature (Quaternary) Phase Shift Keying (QPSK): A modulation technique that has four possible states for each transmitted symbol. Provides two bits per symbol.

rain model: A model that predicts the amount of rain in various parts of the world. Important because K_u and K_a bands have significant rain attenuation.

raised cosine filter: The shape of the base band filter used to reduce the sidelobes of the QPSK signal.

random access: The process of beginning to read and encode the coded bitstream at an arbitrary point.

Reed-Solomon Code: A nonbinary block code frequently used as the outer code in a concatenated coding system.

Regional Administrative Radio Council (RARC): A regional version of WARC.

renewability: Easily changing a CA system to plug security breaches or to introduce technological advances.

repudiation: Denial that a specific event has taken place. In a broadband network, the event could be an order for products or services, or for their cancellation.

reserved: When used in the clauses defining the coded bitstream, indicates that the value may be used in the future for ISO-defined extensions. All reserved bits shall be set to '1'.

S-Curve: The power-out versus power-in curve for a TWT-A or other power amplifier.

secret key: The key that encrypts and decrypts in symmetric algorithms. To prevent interception by unauthorized users, secret keys must be distributed by secure means. Secret keys are good for high-speed data streams, such as content (service) scrambling. DES and DVB Superscrambling are popular secret-key algorithms.

shaped reflector antenna: A specially shaped antenna that transmits more energy into one area than others. This is usually done to normalize the effects of rain.

signal security: The technology used to prevent reception of a service without authorization.

smart card: An insertable credit card–sized device with imbedded processor(s) that provide EMM and ECM functions and a means of secure electronic storage. A smart card can be programmed to decrypt messages, verify messages and digital signatures, and create digital signatures for outgoing messages. Most smart-card implementations are based on the ISO Standard 7816.

source stream: A single nonmultiplexed stream of samples before compression coding.

start codes [system]: Unique 32-bit codes that are embedded in the coded bitstream. They are used for several purposes, including identifying some of the layers in the coding syntax. Start codes consist of a 24-bit prefix (0x0000001) and an 8-bit **stream_id** as discussed in Chapter 7.

STD input buffer [system]: A first-in first-out buffer at the input of system target decoder for storage of compressed data from Elementary Streams before decoding.

still picture: A coded still picture consists of a video sequence containing one coded picture that is intracoded. This picture has an associated PTS. The presentation time of succeeding pictures, if any, is later than that of the still picture by at least two picture periods.

SYNCOM: The first geosynchronous communication satellite. It was developed by Hughes Aircraft Company for NASA.

System Clock Reference (SCR) [system]: A time stamp in the Program Stream from which decoder timing is derived.

system header [system]: A data structure that carries information summarizing the system characteristics of the multiplexed stream.

system noise temperature: The effective temperature of the receiver that determines the noise part of the signal-to-noise ratio (SNR).

system target decoder (STD) [system]: A hypothetical reference model of a decoding process used to describe the semantics of a multiplexed bitstream.

Time Division Multiplex (TDM): The combining of multiple services into a single bitstream.

Time stamp [system]: A term that indicates the time of an event.

transponder: The subsystem on a satellite that receives the Earth-transmitted signal, translates its frequency, amplifies it, and retransmits it to Earth.

Transport Stream packet header [system]: A data structure used to convey information about the Transport Stream payload.

triple-DES: A security enhancement of single-DES encryption that employs three successive processing-DES block operations. Different versions use either two or three unique DES keys. This enhancement is considered to be highly invulnerable to all known cryptographic attacks.

Tuner: The part of an IRD that selects a particular transponder.

variable bit rate: An operation in which the bit rate varies with time during the decoding of a compressed bitstream.

World Administrative Radio Council (WARC): Quadrennial meetings held by the ITU that establish world communication regulations.

XPD (Cross-Polarization Discrimination): The discrimination against the undesired signal of the opposite polarization than the one desired.

INDEX

A

B

C

D

M

N

O

Q

R

S

WXYZ

The Addison-Wesley Wireless Communications Series

Dr. Andrew J. Viterbi, Consulting Editor
http://www.awl.com/cseng/wirelessseries/

0-201-63374-4
272 pages
Hardcover

CDMA
Principles of Spread Spectrum Communication
by Andrew J. Viterbi

CDMA, a wireless standard, is a leading technology for relieving spectrum congestion caused by the explosion in popularity of wireless communications. This book presents the fundamentals of CDMA so that you can develop systems, products, and services for this demanding market.

Andrew J. Viterbi is a pioneer of wireless digital communications technology. He is best known as the creator of the digital decoding technique used in direct broadcast satellite television receivers and in wireless cellular telephones, as well as numerous other applications. He is co-founder, Chief Technical Officer, and Vice Chairman of QUALCOMM Incorporated, a developer of mobile satellite and wireless land communication systems employing CDMA technology.

0-201-63469-4
288 pages
Hardcover

Mobile IP
Design Principles and Practices
by Charles E. Perkins

This book introduces the TCP/IP savvy reader to the design and implementation of Internet protocols that are useful for maintaining network connections while moving from place to place. Written by Charles E. Perkins, a leader in the networking field, this book addresses Mobile IP, route optimization, IP version 6, dynamic Host Configuration Protocol (DHCP), encapsulation, and source routing.

Charles Perkins is currently a senior staff engineer at Sun Microsystems, Inc. Previously he was associated with the IBM T. J. Watson Research Center. Mr. Perkins is a recognized leader and innovator in the field of mobile computing. He is the document editor for the Mobile IP working group of the Internet Engineering Task Force (IETF). Charles Perkins is also is a co-author of DHCPv6 (version 6) and is an Internet Architecture Board (IAB) member. He is a three-time recipient of IBM's Invention Achievement Award.

0-201-63394-9
544 pages
Hardcover

Wireless Multimedia Communications
Networking Video, Voice, and Data
by Ellen Kayata Wesel

This book is a comprehensive guide to understanding the design of wireless multimedia communication systems. *Wireless* is synonymous with *mobile*, enabling the computer user to remain connected while moving from one place to another. *Multimedia* denotes a mix of video, voice, and data information—each of which has different transfer requirements. The author has made significant contributions to the theory, design, standardization, and policies of several wireless multimedia systems. The book is comprehensive in scope, and addresses in detail each of the key elements of a complete wireless multimedia system, including propagation characteristics, modulation, intersymbol interface mitigation, coding, medium access protocols, and spectrum and standards networking, while defining their relationship to one another.

Dr. Ellen Kayata Wesel is a senior scientist at Hughes Communications, Inc. (HCI), where she designs the next generation of wireless multimedia communication systems. Prior to joining HCI, she designed the physical and data link layers for high data rate wireless LANs at Apple Computer.

The Addison-Wesley Wireless Communications Series

Dr. Andrew J. Viterbi, Consulting Editor
http://www.awl.com/cseng/wirelessseries/

0-201-63470-8
448 pages
Hardcover

Wireless Personal Communications Systems
by David J. Goodman

Goodman presents the technology and underlying principles of wireless communications systems. He describes nine important systems: AMPS, IS-41, NA-TDMA, CDMA, GSM, CT-2, DECT, PHS, and PACS. Each system is described using a unified framework so that the reader can easily compare and contrast the systems. Key features, such as architecture, radio transmission, logical channels, messages, mobility management, security, power control, and handoff, are addressed. An analysis of design goals—low price, wide geographical coverage, transmission quality, privacy, and spectrum efficiency—helps the reader to understand why the various systems have such divergent designs.

David J. Goodman is the director of the Wireless Information Networks LABoratory (WINLAB) and is a member of the faculty at Rutgers University. Previously, Dr. Goodman worked at Bell Laboratories for twenty-one years, where he did pioneering research in wireless communications.

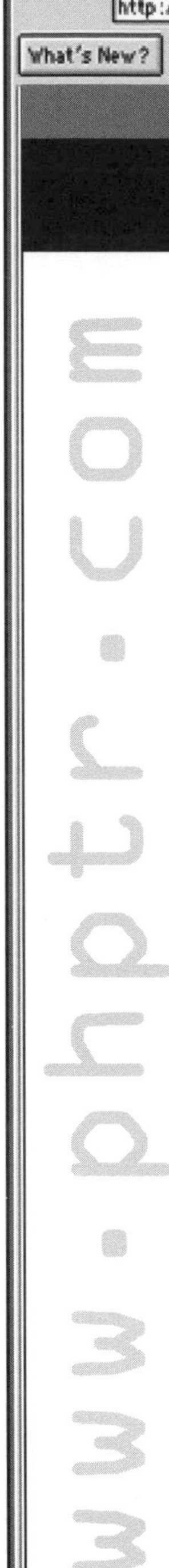

Keep Up-to-Date with

PH PTR Online!

We strive to stay on the cutting-edge of what's happening in professional computer science and engineering. Here's a bit of what you'll find when you stop by **www.phptr.com**:

- **Special interest areas** offering our latest books, book series, software, features of the month, related links and other useful information to help you get the job done.
- **Deals, deals, deals!** Come to our promotions section for the latest bargains offered to you exclusively from our retailers.
- **Need to find a bookstore?** Chances are, there's a bookseller near you that carries a broad selection of PTR titles. Locate a Magnet bookstore near you at www.phptr.com.
- **What's New at PH PTR?** We don't just publish books for the professional community, we're a part of it. Check out our convention schedule, join an author chat, get the latest reviews and press releases on topics of interest to you.
- **Subscribe Today!** **Join PH PTR's monthly email newsletter!**

Want to be kept up-to-date on your area of interest? Choose a targeted category on our website, and we'll keep you informed of the latest PH PTR products, author events, reviews and conferences in your interest area.

Visit our mailroom to subscribe today! **http://www.phptr.com/mail_lists**